Protein Crystallography

Protein Crystallography

A Concise Guide

Eaton E. Lattman and Patrick J. Loll

The Johns Hopkins University Press

Baltimore

© 2008 The Johns Hopkins University Press

All rights reserved. Published 2008

Printed in the United States of America on acid-free paper

9 8 7 6 5 4 3 2 1

The John Hopkins University Press

2715 North Charles Street

Baltimore, Maryland 21218-4363

www.press.jhu.edu

Lattman, Eaton.

Protein crystallography : a concise guide / by Eaton E. Lattman and Patrick J. Loll.

p. ; cm.

Includes bibliographical references and index.

ISBN-13: 978-0-8018-8806-9 (hardcover : alk. paper)

ISBN-10: 0-8018-8806-9 (hardcover : alk. paper)

ISBN-13: 978-0-8018-8808-3 (pbk. : alk. paper)

ISBN-10: 0-8018-8808-5 (pbk. : alk. paper)

1. Proteins—Structure. 2. X-ray crystallography. I. Loll, Patrick. II. Title.

[DNLM: 1. Proteins—chemistry. 2. Crystallography, X-Ray—methods. 3. Protein Conformation. 4. Proteomics—methods. QU 55 L364p 2008]

QP551.L345 2008

612'.01575—dc22   2007036099

A catalog record for this book is available from the British Library.

Special discounts are available for bulk purchases of this book. For more information, please contact Special Sales at 410-516-6936 or specialsales@press.jhu.edu.

The Johns Hopkins University Press uses environmentally friendly book materials, including recycled text paper that is composed of at least 30 percent post-consumer waste, whenever possible. All of our book papers are acid-free, and our jackets and covers are printed on paper with recycled content.

To David Sayre

Contents

Preface

Why another book on protein crystallography? Because we believe that there is a demand for a concise, accessible introduction to this powerful method. Proteins, as the agents that carry out the functions of cells and tissues, are assuming a paramount role in research areas ranging from drug development to cell biology. The atomic structures of proteins, as revealed by crystallography, are increasingly being used to provide detailed insights into function and mechanism. Indeed, dealing with protein structure is becoming unavoidable. Thus, the number of scientists who desire a basic grasp of crystallography is growing, but there has not been a volume available that offers an appropriate introduction.

We have designed our book to meet this need. First, it is quite short, less than half the length of most other books on this topic. Second, we have worked hard to make it accessible. It has lots of illustrations, and equations are introduced by using intuitive arguments. Third, it focuses tightly on crystallography as an imaging or microscopical process, deemphasizing the practical details. Finally, it contains a glossary that defines many crystallographic terms. The book is aimed at anyone who would like a trip through the essentials of crystallography. It will not turn you into a crystallographer; but we hope reading it will make it easier to converse with one.

The authors are grateful to Alexander McPherson and Jason McLellan for the gift of figures and to Michaelis Hadjithomas for creating and drawing many of the figures in the book. They thank each other for support during periods of slow progress. Many figures in this book were prepared with the help of PyMOL (De-Lano, W. L. The PyMOL Molecular Graphics System [2002]; www.pymol.org).

Protein Crystallography

Introduction

1.1 What Is X-ray Crystallography?

In 1912 Walter Friedrich and Paul Knipping, following up on Max von Laue's prediction, illuminated a crystal of copper sulfate with X rays and obtained the first X-ray diffraction pattern. This simple experiment has evolved into a powerful technique that has enabled us to decipher the structure of DNA and of protein molecules. This technique has resulted in at least eight Nobel prizes, and researchers in the field anticipate more to come.

X-ray crystallography is a powerful form of microscopy that allows us to visualize atoms and molecules. Almost all that we know about the three-dimensional structures of proteins and nucleic acids, we have learned from the use of X-ray crystallography. This method provides images of molecular structures that are far more detailed than any provided by light microscopy because the very short wavelengths of X rays allow them to "sense" structural variation at the atomic level.

X-ray crystallography differs from the more familiar light microscopy in several important ways. The key differences are:

1. Crystallography uses X rays instead of visible light. The X rays used in crystallography have wavelengths of about $1 \text{ Å} = 10^{-10}$ meters, while visible light has a wavelength of $\sim 5000 \text{ Å}$.
2. Unlike microscopy, X-ray crystallography is lensless. We are familiar with the idea that X rays pass straight through objects—like people!—without being

deflected. This suggests why there are no substances that make good lenses for X rays: it is the job of a lens to bend or refract light, and X rays are so energetic that they resist bending.

3. The specimens used in crystallography are crystals, which contain many perfectly repeated copies of the molecules we wish to see. In contrast, microscopists frequently examine single objects.

The very short wavelength of X rays is what makes them useful for studying the structure of matter at the atomic level. The ability of a microscope to resolve

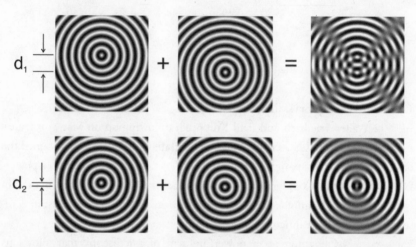

Figure 1.1. Summation of two circular waves emanating from two sources near to each other. Such waves might be produced by light interacting with two small holes in an optical mask, or by two pebbles dropped simultaneously into a pool of water. Bright areas represent the wave crests, and dark areas correspond to the wave troughs. The two individual waves are shown on the left, and their sum is shown on the right. In the upper panels, the waves are emanating from points that are separated by a distance $d_1 \approx$ twice the wavelength. The resultant wave shows clear interference patterns, which would reveal to an observer some distance away that this resultant wave pattern contains contributions from two waves. In other words, the observer can *resolve* the two sources. In the lower panels, the waves emanate from points separated by a distance d_2, which is less than the wavelength. In this case, the resultant wave pattern appears almost identical with a wave coming from a single source. It is impossible to determine whether this wave pattern was generated from one pebble or two, and an observer would therefore be unable to resolve the two sources. In this manner, the wavelength of the scattered radiation limits our ability to distinguish two closely spaced points.

fine details is limited by the wavelength of the radiation used. For example, two carbon atoms joined by a single bond are about 1.5 Å apart. To resolve these two atoms as separate objects, we need to use a wavelength of light that is (roughly) no larger than the distance between the atoms; light with such a short wavelength falls in the X-ray region of the spectrum.

We give the name *resolution* to the ability to distinguish or resolve two nearby points. The effect of wavelength upon resolution is illustrated in Figure 1.1. The upper panel shows waves emanating from two well-separated points; the lower panel shows what happens when the two points are separated by less than one wavelength. In the latter case, the two wave patterns blend together and can hardly be distinguished from the waves emanating from a single point. Thus, it is impossible to know whether the pattern of scattered waves results from two closely spaced objects or a single larger object.

We have described X rays as electromagnetic waves with short wavelengths. It is one of the counterintuitive findings of quantum mechanics that light can be equally well described as a stream of particles called *photons*. The energy of a photon of light is inversely proportional to its wavelength. Because the X-ray photons used in crystallography have short wavelengths, they have high energies (~10,000 electron volts). The highly energetic character of X rays complicates the imaging process, as described below.

In an ordinary light microscope, light waves scattered by the specimen are collected by a high-quality lens (or system of lenses) and focused to form an image. This is illustrated in Figure 1.2a. Because X rays have much greater penetrating power than visible light, it is more difficult to fabricate lenses that are able to focus X rays. The best X-ray lenses currently known are of poor quality compared with optical lenses, and they cannot produce high-resolution images. Thus, as shown in Figure 1.2b, we carry out crystallographic experiments without lenses, collecting the scattered radiation directly on film or some other detector. The pattern that the scattered radiation makes on the detector is the *diffraction pattern*. We use this measured diffraction to perform calculations on a computer that mimic the action of the lens. These calculations allow us to recombine the scattered radiation to form an image.

Unfortunately, calculating the image is not quite as simple as we have suggested. Experimental limitations preclude our extracting all the information contained in the diffraction pattern. Specifically, a complex number is required to completely describe the scattered waves that make up the diffraction pattern;

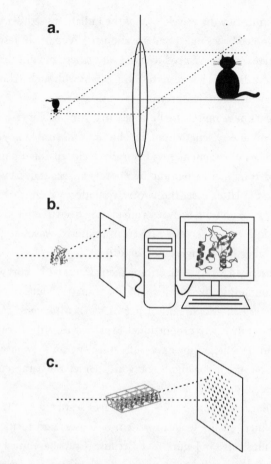

Figure 1.2. (a) Schematic of a conventional imaging experiment with visible light. Here, the specimen is represented by a cat. Radiation scattered by the cat is collected by the lens, and an image is created at the image plane of the lens. For each point in the specimen, the lens collects light scattered in many different directions and redirects it to form an image of that point. (b, c) Schematics of imaging experiments involving proteins. In (b) a single molecule is shown scattering X rays that are collected by a detector with no intervening lens. The pattern on the detector will look nothing like the object itself. The computer and monitor symbolize the process by which calculations that mimic the function of a lens are used to create an image. In (c) a protein crystal is shown as the specimen, along with the punctate diffraction pattern characteristic of crystals. In (b) and (c) the horizontal line entering from the left represents the incident X-ray beam and the diagonal lines represent diffracted rays. Adapted from *Crystallography Made Crystal Clear* by Gale Rhodes.

however, we are able to measure only the squared modulus of this complex number. We need to know the correct complex square root of this number to create the image. In practical terms, we can measure the amplitude of the diffracted rays but not their relative phases. The problem of recovering the correct square roots is the *phase problem* and is considered in more detail in Chapter 3.

Crystallographic experiments require crystals, as shown in Figure 1.2c. Why do we use crystals? In principle we could do an X-ray experiment to image a single molecule, but there are two practical obstacles to this. First, it would be impossible to measure the diffraction pattern from a single molecule because it would be too weak and drowned in noise from scattering by other elements of the system. The second obstacle is specimen damage; a single molecule would be burned up by the X rays before it could give rise to a useful diffraction pattern. Crystals help us with both of these difficulties.

Crystalline specimens greatly increase signal-to-noise in the measured diffraction pattern. The protein* crystals we use contain 10^{12} or more molecules, and so diffract at least 10^{12} times as much radiation as a single molecule. For reasons that we discuss later, the diffraction pattern from a crystal is confined to rays or beams emanating in certain directions, which form "spots" on the film, making the pattern much easier to measure. These spots are often called *reflections* because the incoming X-ray beam appears to be reflected by the crystal to form the outgoing ray.

A typical example of a crystallographic diffraction pattern is shown in Figure 1.3. The complete diffraction pattern comprises many pictures such as this one, taken with the crystal in various orientations. The number of independent reflections in the diffraction pattern of a protein crystal is very large, typically tens or even hundreds of thousands. This is a tremendous amount of information, which should not be surprising—the diffraction pattern specifies the detailed image of a molecule that may contain thousands of atoms.

Crystalline specimens also reduce specimen damage. The X rays that create the diffraction pattern are scattered from all the molecules in the crystal lattice, so each individual molecule receives a much smaller dose. For example, if 10^{12} photons are required to create a high-resolution diffraction pattern and a crystal contains 10^{12} molecules, then on average each molecule is hit by only one photon.

---

*This book describes the use of crystallography to determine protein structure. The reader should be aware, however, that most of the techniques we discuss are equally applicable to other macromolecules such as DNA.

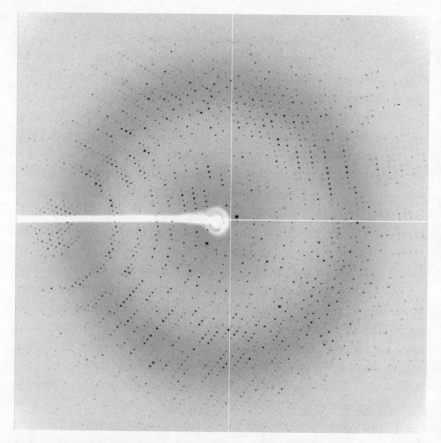

Figure 1.3. This figure shows the diffraction pattern created when a crystal of the protein ferritin was rotated through a small angle (0.5°) while illuminated by the X-ray beam. The white feature extending from the left is the shadow of the beam stop, a small piece of metal positioned to capture the undiffracted X rays that pass through the crystal. The reasons behind the punctate character of the diffraction pattern will be discussed later in the book.

Crystallography is an imaging technique. Any image maps a particular characteristic of an object—color, reflectivity, and so on. X rays are scattered by electrons, and so our image is a record of the spatial distribution of the molecule's electrons. This image is the *electron density function*. The maximum resolution that can be achieved in the electron density image is determined by the wavelength of the light used. However, factors such as disorder in the crystals degrade the resolution of the image, and as a result it is generally not possible to

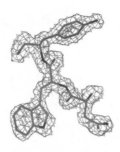

Figure 1.4. This stereo pair illustrates the results of a crystallographic experiment and the interpretation of those results. The wire mesh contours show the experimentally determined electron density function. The stick model shows how this density is interpreted in terms of an atomic model. See "Viewing Stereo Pictures" for more information.

resolve individual atoms in protein structures. Figure 1.4 shows a portion of the electron density map for a protein molecule, presented as a stereo view.

To understand how to visualize electron density maps, it is useful to consider a simpler example, such as the weather map shown in Figure 1.5. A weather map is a means of visualizing a two-dimensional function, namely, atmospheric pressure presented as a function of latitude and longitude. Points having the same atmospheric pressure are linked by lines on the weather map called *isocontours*. In a three-dimensional function such as the electron density function, isocontours will take the form of surfaces. These surfaces are frequently drawn to resemble a wire mesh (see Figure 1.4). This representation has the advantage of being transparent; it allows crystallographers to see the interior of a complex molecule. Drawing multiple contours in a three-dimensional electron density map would give rise to nested sets of surfaces and produce images that are dense and difficult to understand. Therefore (unlike the case with two-dimensional maps), electron density maps in general are visualized by showing only one contour at a time.

Electron density maps provide interesting pictures of molecules, but what we really want from a crystallographic experiment is a quantitative representation of a molecule's atomic structure. We want to know exactly where all the atoms are— to develop a list of the x, y, z coordinates of each atom in the molecule. From such a list we can calculate many things: the distances between atoms and the angles between bonds, for instance. To obtain these coordinates, the experimental electron density contours are superimposed on a stick drawing representing the

Viewing Stereo Pictures

Protein structure figures are commonly published as stereo pairs. We owe our ability to see objects in three dimensions to the fact that our eyes are separated from each other by about 6 cm. Each eye therefore sees an object from a slightly different perspective, and our brains can integrate these different perspectives to produce an impression of the object's three-dimensional structure. Images that utilize stereo pairs take advantage of this ability by providing two slightly different views of the molecule, corresponding to what would be seen by each of the two eyes. By looking at the left-hand panel with the left eye and the right-hand panel with the right eye, readers can gain a very impressive and helpful sense of depth. It takes some effort to learn to do this, but it is well worth it for anyone interested in molecular structure. A complete and useful introduction to stereo viewing is given on Gale Rhodes' Web site at: www.usm.maine.edu/~rhodes/0Help/StereoView.html.

The trick to viewing stereo figures is to convince each of your eyes to look only at its intended panel. We here mention just two approaches to accomplish this. The first is to buy a pair of special glasses that make the process very easy. Searching on the Web for "stereo glasses" or "stereo viewer" will bring up numerous sources. The cheapest glasses have cardboard frames and plastic lenses and cost only a few dollars. A more elaborate pair will set you back more (~$20), but will still be reasonable.

The second approach is to learn to view stereo figures without special tools. This takes some practice, but makes your life much simpler. A method that works for most people is the following: Touch your nose to the page between the two views. Try to relax your eyes. You will see a blurry image. Slowly move the paper away from your eyes, while keeping your eyes relaxed. You will start to see multiple images, but concentrate on the central image. When the picture is in focus, the central image is the fused stereo pair and should show the desired three-dimensional effect. *Note:* If you wear glasses, you will probably find it easier to perform this maneuver without them.

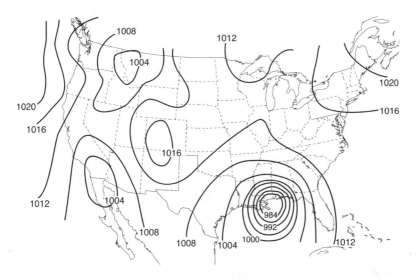

Figure 1.5. Weather map of the continental United States, showing atmospheric pressure isocontours on the day of Hurricane Katrina. Note the steep pressure minimum near New Orleans. In this two-dimensional contour map, isocontour lines represent points at the same pressure; the numbers give the pressure values of different isocontours in units of millibars.

underlying molecular structure. The stick drawing is constructed using computer software that allows us to manipulate the positions of the atoms until they fit well with the experimental image. The model atomic coordinates are then recorded from the stick model.

The process of going from the contour image to a set of atomic coordinates is called *fitting the map*. In Figure 1.4 one can see that the sticks are compatible with the contours, but it is not trivial to go from the contours to the stick model. This process is described in more detail in Section 4.3. Great strides have been made in automating the process of fitting the map, but as of this writing the majority of crystallographic structures cannot be completely fit in an automated manner. The electron density images in these cases are noisy and/or lack resolution, and the correct fit is not necessarily obvious. Fitting such maps requires human judgment. Thus, the initial stick model is an interpretation of the electron density image that is guided by what we know about the structure of proteins.

How do we know how accurate our atomic model is? Once we have a model that specifies where each atom is, we can calculate the diffraction pattern that the crystal would give if the model were correct and compare it with the observed data. The statistic that describes the comparison between calculated and observed is called the R value or R index. The R value provides an objective measure of how well our model agrees with the experimental data. Atomic models should also be consistent with known biochemistry. Thus, for example, groups known to be in the active site of an enzyme should be clustered together in the model.

Sometimes portions of the protein simply cannot be seen in the electron density map. To understand why, it is important to recognize two things. First, the diffraction pattern is averaged over time. The molecules in a crystal are not stationary during the diffraction experiment. The atoms in a protein molecule have the same energies as gas atoms at the same temperature, which means that they are not at rest but are rattling around in a cage formed by their neighbors. These motions are fast relative to the timescale of the diffraction experiment. This will blur the appearance of the atoms in the final image, for the same reason that rapidly moving objects appear blurred in photographs. Second, the diffraction pattern is averaged over space. The diffraction we measure is an average of the scattering from all the molecules in the crystal. If conformational differences exist between different molecules in the crystal, the resulting image will reflect the average of all those different conformations. Thus, when the electron density for a portion of a molecule is poorly defined, it often means that there are many possible conformations or that the molecule is undergoing large motions at that locus.

Remember that the initial stick model is built by a human being and therefore reflects certain subjective decisions on the part of that person. How can we trust the efforts of one person to fit a model to a set of contours? Another person would certainly do it a little differently. Initial models therefore must contain errors or inaccuracies. What rescues crystallographers here is a process called *refinement*. Refinement is a computational procedure that systematically alters the model to maximize agreement between the observed and calculated diffraction patterns. In general, this procedure is very successful in objectifying the model. As a result, when similar or identical structures are solved independently in different laboratories, the resultant atomic models usually agree within experimental error. In the best cases (salts and small organic molecules), crystallography can yield molecular models with accuracies of 0.001 Å in atomic positions. Protein crystals are less well ordered, and the accuracy of protein structures is usually a few tenths of an angstrom. We discuss refinement in detail later in the book.

Finally, refined models are deposited in the Protein Data Bank. Some years

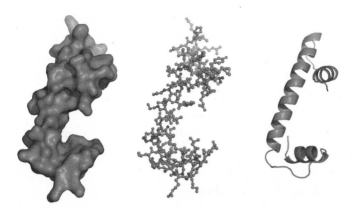

Figure 1.6. Three representations of a portion of a protein structure determined by using an X-ray crystallography experiment. In all cases the underlying information is a computer file containing the x, y, z coordinates of each atom in the structure. The left-hand panel shows a surface representation of the molecule, emphasizing the portions of the protein accessible to the surrounding solvent. The central panel is an all-atom representation (also known as ball-and-stick) in which the atoms are shown as spheres, connected by sticks representing bonds. In this diagram the atomic spheres are drawn much smaller than their true size so that one can see the interior of the molecule. All-atom renditions are most commonly used to illustrate small portions of a molecule; they can be confusing when too many atoms are included. The right-hand panel is a backbone trace, with the flat surface of the ribbon roughly corresponding to the plane of the peptide bonds. This view clearly illustrates that the domain shown comprises three α-helices connected by linker regions. Because of their simplicity, ribbon diagrams are popular choices to illustrate overall topology and are particularly useful to highlight regions of different secondary structure.

ago crystallographers agreed (some reluctantly) to share their results in an open and usable way, by deposition in a public data bank. Today, the Protein Data Bank (www.pdb.org) contains the atomic coordinates of almost all the proteins whose X-ray structures have been determined, almost 50,000 at this writing. A similar database, the Nucleic Acid Database (ndbserver.rutgers.edu), exists for nucleic acid structures. Visit these sites to see your favorite macromolecules.

Atomic models of proteins and other macromolecules are extremely complex, which can make them difficult to comprehend. To address this problem, many different methods have evolved for representing molecular structure. These methods are complementary, in that they highlight different aspects of the structure. Figure 1.6 shows three of the many options available for molecular drawings.

We conclude this section with a summary of the steps needed to complete the X-ray crystallographic study of a protein.

1. Grow crystals of the protein.
2. Measure the diffraction pattern.
3. Estimate phases and form an image = calculate electron density function.
4. Fit a model to the image using the known amino acid sequence.
5. Refine the model.
6. Analyze the model in terms of the molecule's known biochemical functions. Is it consistent? Does it explain things?
7. Deposit coordinates in the PDB.

1.2 A Quick Look at Protein Crystals

Crystals of proteins look a lot like the ones you grew in elementary school science class. Unlike salts and small molecules, however, proteins are large, floppy, and irregularly shaped, and it is often tricky to induce them to form an ordered crystal lattice. There is an art to growing these crystals, which we will not discuss. However, growing crystals usually represents the most difficult and time-consuming part of a macromolecular crystallography project. Because of the large size of protein molecules, there are substantial voids between them in the crystal lattice. The space between proteins in the lattice is not a vacuum; it is filled with the solution from which the crystal was grown. Protein crystals typically contain ~50% solvent by volume, and sometimes as much as 75–80%. These liquid channels often allow small compounds such as ligands to be diffused into an existing crystal. A gallery of protein crystals is shown in Figure 1.7.

The external regularity of crystals mirrors their microscopic order. Crystals are built of parallelepipeds (shoeboxes) called *unit cells*, which are stacked side by side in three dimensions to give a regular array. Each unit cell has a small number of molecules in it, usually related to each other by symmetry operations, such as rotations or translations. The smallest unit needed to build up the whole crystal by repeated symmetry is called the *asymmetric unit*. See Figure 1.8 for a schematic view of how molecules are combined to build up a macroscopic crystal. The study of crystal symmetry is one of the most historically important areas of crystallography. In the interest of conserving space, we will give it only a very brief treatment; see the text by Stout and Jensen if you would like to know more.

Crystals may possess three basic types of symmetry: rotations, mirrors and

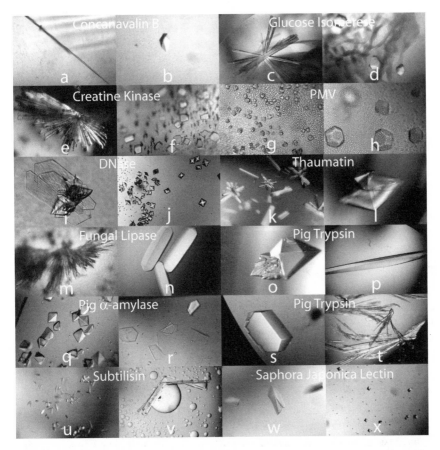

Figure 1.7. Gallery of pictures illustrating a variety of protein crystals. Note the wide variety of shapes. Note also the variation in the perfection of the external appearance of the crystals, which is not necessarily related to perfection in internal order. The names of the proteins appear in the individual frames. Kindly provided by Alexander McPherson.

inversions, and translations. In a crystal possessing *rotational symmetry*, every molecule in the crystal is superimposed on an identical copy of itself when rotated by a specific angle (for example, 180°) about a particular axis. Allowed rotational symmetries are twofold (180°), threefold (120°), fourfold (90°), and sixfold (60°). Note that fivefold symmetry is not allowed in crystals, nor is sevenfold symmetry or higher. When we say they are not allowed, we mean that it is physically impossible to build up a repeating three-dimensional array that is based on fivefold or sevenfold symmetry. *Mirror symmetry* or *inversion symmetry*

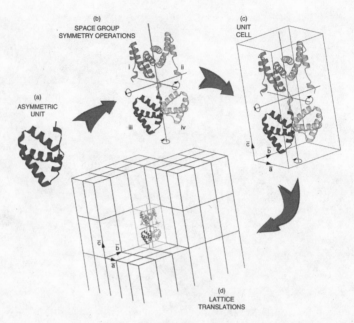

Figure 1.8. Illustration of how a crystal is built up by symmetric repetition of simple elements. (a) The *asymmetric unit* is the smallest entity that is necessary to build up the entire crystal. In this example, the asymmetric unit corresponds to a single molecule. (b) Identical copies of this molecule are generated by the space group symmetry operations. In the example shown, each of the four molecules in the unit cell is related to the other three by twofold (180°) rotations about one of three symmetry axes. The three rotational symmetry axes are parallel to the unit cell edges. This type of packing arrangement is known as 222 symmetry. These four molecules comprise the contents of the unit cell, which is shown in (c). The unit cell is a box that encloses the various symmetry-related copies of the asymmetric unit. The edges of the unit cell are defined by three vectors, *a*, *b*, and *c*. Finally, as shown in (d), multiple copies of the unit cells are stacked together to form the crystal, much as bricks are stacked to form a wall. Each unit cell is related to all of its neighbors by a pure translation that constitutes an integer number of steps in *a*, *b*, and *c*. Kindly provided by Alexander McPherson.

means that molecules in the crystal are superimposed on copies of themselves when reflected through a particular plane or point. Mirror planes and inversions change the hand of objects and can therefore not be present in protein crystals, since the amino acids comprising proteins are chiral. Finally, *translations* can be combined with rotations or mirror planes to give screw axes or glide planes, respectively.

Crystal lattices are divided into seven types, depending on the rotational symmetry they contain. Cubic lattices are the most symmetric, containing four-fold, threefold, and twofold rotation axes (432 symmetry). Tetragonal lattices have fourfold symmetry, hexagonal lattices sixfold, and so on. Triclinic lattices have no rotational symmetry. The shape of the unit cell—the lengths of the edges and the angles between them—depends on the lattice symmetry. For example, in triclinic lattices, which are the least symmetric, none of the unit cell edges is equal to any other, and none of the angles is 90° or 120°; in tetragonal lattices two sides of the unit cell are equal, and all angles are 90°; and so on.

Exhaustive calculations have proven that exactly 230 different possible combinations of lattices and symmetries exist; in other words, there are 230 (and only 230) distinct ways to pack identical copies of objects into a three-dimensional lattice. Each of these packing arrangements is called a *space group*. Here, the word group refers to an algebraic structure in which operators (symmetries) operate on some element to produce another. Of the 230 space groups, only 65 are possible for proteins: the rest involve mirror symmetries that would change the hand of the molecule. (Note that it *is* possible to produce crystals of chiral molecules that contain mirror symmetry, if one crystallizes the racemic mixture.)

The precise space group in which any protein will choose to crystallize is impossible to predict. And to an experimenter who is mainly interested in the structure and biological function of a protein, the details of the crystal symmetry can seem artificial and irrelevant. However, because crystallography involves creating an image of the contents of the crystal, the exact manner in which those contents are arranged turns out to be important, and knowledge of the space group is a prerequisite for structure determination. Fortunately, the symmetry of the crystal lattice is reflected in the symmetry of the diffraction pattern. For example, if a crystal is oriented so that a fourfold rotational symmetry axis within the crystal is parallel to the X-ray beam, the diffraction pattern will show fourfold symmetry. Thus, careful analysis of the diffraction pattern allows the determination of the space group. This determination can generally be carried out automatically by the software that processes the diffraction data.

1.3 Noncrystalline Specimens

Crystals offer important advantages in diffraction experiments, and the most detailed structural information available is derived from scattering by crystals, but useful structural information can also be obtained from noncrystalline specimens. Historically the most important noncrystalline diffraction specimens have been fibers. Fibers are built from linear structures that repeat in one dimension but not the other two—one-dimensional crystals, in effect. DNA is a famous example, and α-helices are another. In both of these cases the repeated element is small—a base pair or an amino acid. In some fibers, such as F-actin, the repeating unit is an entire protein molecule. A macroscopic fiber is composed of many of these units, with their long axes approximately parallel. However, these units are *randomly* oriented about their long axes, so that the fiber is a rotationally averaged structure. Fibers are therefore much less ordered than crystals, and their diffraction patterns contain correspondingly less information.

When the repeating unit is small, the diffraction from helical structures can often be used to deduce an atomic model of the structure, much as Watson and Crick did for DNA. The process does not usually involve generating an image by solving the phase problem; rather, a trial-and-error method is used in which a model is built and the model's calculated diffraction pattern compared with the observed. When the repeated unit is a whole protein, the lack of information in the diffraction pattern prevents an atomic model from being built unless the structure of the protein monomer is known independently. A notable exception to this rule is the fiber formed by the tobacco mosaic virus—the structure of this virus was determined solely from the fiber diffraction pattern.

Molecules in solution are even less well ordered than those in a fiber. In a solution at any given instant, we expect the molecules to be found in all possible orientations. Besides, unlike a fiber or a crystal, in solution the molecules are free to move, and so their orientations will change from one moment to the next. For this reason, the diffraction pattern measured from a solution is the rotational average of the pattern from a single molecule. Solution scattering experiments were historically carried out by measuring radiation scattered within a small angle of the incident beam, and so the method has been called small-angle X-ray scattering or SAXS. However, recent instrumental advances have enabled us to measure radiation scattered at larger angles as well. The principal kinds of results that can be derived from solution scattering experiments are summarized below.

- From scattering very near the incident beam we can find molecular weight.
- From scattering at small angles we can determine molecular size, represented by the *radius of gyration*, R_G.
- At larger angles we can determine the approximate shape of the molecule. This allows us to distinguish between a folded and an unfolded protein, for example.

- New intense X-ray sources have enabled time-resolved solution scattering experiments. By monitoring changes in size and shape over time, investigators can follow processes such as protein folding and oligomerization.

1.4 Summary

- X-ray crystallography is an imaging technique, like microscopy. It allows us to directly image the molecules that make up crystals.
- X-ray crystallography is lensless. Because X rays have very high energies, we are unable to fashion lenses capable of bending them. Because we lack good lenses, we must mimic the action of a lens by measuring the diffracted radiation and using it to calculate images.
- X rays have wavelengths comparable to the distances between atoms in molecules. This means that individual atoms can be resolved in images of molecules.
- Crystals are composed of a set of shoebox-like unit cells arrayed on a periodic, three-dimensional lattice. The densely packed and well-ordered arrays of molecules found in crystals give rise to strong scattering signals, with much higher signal-to-noise ratios than individual molecules would give. Also, the X-ray scattering necessary to produce an image can be distributed over all the unit cells in a crystal, reducing specimen damage.
- Noncrystalline samples such as fibers and solutions are also useful specimens for X-ray diffraction.

FURTHER READING

Several excellent texts cover topics discussed in this chapter, approaching the topics from varied points of view.

Crystallography Made Crystal Clear, 2nd edition, by Gale Rhodes (Academic Press, New York, 2003) is a clear and well-constructed paperback. Good figures; gentle math.

Outline of Crystallography for Biologists by David Blow (Oxford University Press, Oxford, 2002) is written by one of the creators of protein crystallography. Makes extensive use of the convolution operation to explain many crystallographic points. This is a powerful tool, but it comes with a learning curve. Wonderful physical insights.

The Optical Principles of the Diffraction of X-Rays by R. W. James (G. Bell and Sons, Ltd., London, 1954) is a bible for the more physically inclined.

Many of our figures are adapted from *Crystal Structure Analysis: A Primer,* 2nd edition, by J. P. Glusker and K. N. Trueblood (Oxford University Press, New York, 1985). This book is more oriented to small-molecule crystallography, but the early chapters that describe crystal diffraction show great clarity and physical insight. Inexpensive.

X-ray solution scattering is making gigantic strides because of new algorithms for data analysis, and because of better instrumentation. See the outstanding review by Svergun and colleagues: Koch, M. H., Vachette, P., and Svergun, D. I. Small-angle scattering: a view on the properties, structures and structural changes of biological macromolecules in solution. *Q. Rev. Biophys.* 2003;36(2):147–227.

Progress in fiber diffraction is summarized in a fine review by Gerald Stubbs: Stubbs, G. Developments in fiber diffraction. *Curr. Opin. Struct. Biol.* 1999;9(5):615–619, which also provides links to earlier reviews and papers.

The art of protein crystallization is beautifully presented in *Crystallization of Biological Macromolecules* by Alexander McPherson (Cold Spring Harbor Laboratory Press, Cold Spring Harbor, NY, 1999).

See *Introduction to Protein Architecture* by Arthur Lesk (Oxford University Press, Oxford, 2003) for an insightful and elegantly illustrated guide to protein structure.

2

A Physical Understanding of Diffraction

2.1 What Is Diffraction?

Diffraction is the process by which matter deflects light rays from their straight-line path. This process is also called scattering. In microscopy, light scattered by the specimen is captured and focused by lenses to make an image. In crystallography, the light scattered by the sample is captured by a detector (Figure 2.1); we obtain an image by performing calculations that mimic the action of the lens.

The precise pattern made by the scattered light is called the *diffraction pattern*. In general, a diffraction pattern does not resemble the object that gave rise to it, but the appearance of the diffraction pattern is critically dependent on the object's structure. This chapter presents a general method for understanding how an object and its diffraction pattern are related and how an image of an object can be recovered from its diffraction pattern. The discussion is slanted toward crystals and crystallography, but it is generally applicable to many imaging techniques.

Illuminate a simple one- or two-dimensional crystal with a laser pointer, and you will observe the basics of diffraction. The diffraction pattern of such a crystal can be projected onto a wall or screen and appears as a lattice of spots. Some of the questions that might occur to you include the following:

- Why is the diffraction pattern of a crystal confined to spots (called *reflections* by crystallographers)?
- How is the arrangement of reflections related to the crystal lattice?
- How can we use the reflections to make an image of the specimen?

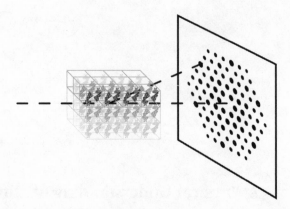

Figure 2.1. Schematic of a protein crystal and its X-ray diffraction pattern. The horizontal line entering from the left represents the X-ray beam used to illuminate the crystal; the square on the right represents the detector. Many different diffracted rays emerge from the crystal and are captured by the detector; one such ray is represented by the diagonal dashed line emerging from the crystal. Note the punctate character of the diffraction pattern on the detector surface, with its regular arrangement of points. Each point represents the spot where a different ray emerging from the crystal strikes the detector.

• How is the diffraction pattern of a single object related to that of a crystal containing many copies of that object?

You should try this yourself. An ideal specimen is one of the grids used to hold samples in the electron microscope; it looks like a piece of window screen, but it has a much finer mesh. If you have access to a local electron microscopy facility, you may be able to beg one. There are also Web sites that allow you to download files containing simple one- and two-dimensional lattices designed to be printed onto a sheet of transparency film.* Simply shine the laser beam through the specimen and see what happens (an EM grid is very small, so you'll need to use a forceps to hold it, or you could affix it to something else using glue or modeling clay). This experiment is shown in Figure 2.2.

---

*As of this writing, you can obtain such files at the following URLs: www.math.montana.edu/frank w/ccp/GraphPaper/diffraction/index.htm and http://alpdmn.phys.psu.edu/gratings/. These links may no longer be valid by the time you read this, but a search with keywords such as "laser printer diffraction grating" and "laser printer diffraction patterns" is likely to yield other possibilities. It is also possible to purchase demonstration gratings from scientific supply houses such as Edmund Scientific. Finally, EM grids can be purchased inexpensively from suppliers on the Web.

Figure 2.2. A simple laser diffraction experiment. In the upper panel, an inexpensive laser pointer (1) is used to illuminate an EM grid (2), which is glued to a paperclip and affixed with modeling clay to an inverted drink cup. The diffraction pattern (3) is projected onto a wall about 1.5 meters away. The EM grid appears as a bright spot because it is reflecting light back toward the camera. The lower panel shows the same setup, except with the room lights turned down so the details of the diffraction pattern may be seen. The inset shows a photomicrograph of the EM grid; the scale bar in the inset corresponds to 500 μm. Note the ripples that appear around the EM grid in the lower panel; these represent interference patterns in the reflected light. (You don't need to clamp the laser pointer to do this experiment yourself; the clamp simply serves to hold it steady so the photograph isn't blurred.)

Before we discuss this further, we need to review a few aspects of the behavior of light. Many different kinds of light (including X rays) can be generated by accelerating electrons. In an incandescent light bulb, for example, the large thermal energies of the atoms in the filament cause their electrons to undergo violent changes in motion, or accelerations. These accelerations cause the electrons to give off light. This process is reciprocal: X rays and other light waves can accelerate electrons, which in turn causes the electrons to emit radiation.

How does light accelerate electrons? Recall that light propagates as an oscillating wave. What is oscillating is the force that the light beam exerts on electrons in its path. If you shine a light wave onto an electron, the electron experiences this oscillating force, called the *electric field*, *E*. The electric field causes the electron to oscillate back and forth with the frequency of the incoming radiation. As the electron oscillates, it is constantly changing velocity, which means it is accelerating. Hence, the electron will reradiate light at the same frequency as the incoming light and with an amplitude proportional to the light's electric field. When an object such as a crystal is illuminated with X rays, each electron in the object oscillates and reradiates X rays. The diffraction pattern seen by a distant observer is the sum of all the X rays being scattered or diffracted by the crystal.

Figure 2.3 shows several examples of adding waves together. Later on we will discuss how to do this quantitatively, but the most important result is very simple. When you add together several cosine waves of the same wavelength, you obtain a resultant cosine wave of the same wavelength. Thus, adding together all the waves scattered by the different electrons in the sample yields a new (resultant) wave.

The model for generating the diffraction pattern of an object by adding the diffracted waves was developed by the Dutch physicist Christiaan Huygens in the seventeenth century. To illustrate, consider dropping a pebble into a pond. It gives off a circular wave or ripple. If you drop two pebbles in at different places you obtain two intersecting sets of ripples (recall Figure 1.1). An observer at the surface of the water sees the sum of the two waves. If a crest from one wave falls on a crest from the other wave, the resultant wave is twice as high; if a crest falls on a trough, the resultant wave vanishes. When a specimen is illuminated by light (or X rays) every electron in it reradiates a spherical wave, analogous to the circular wave caused by the pebbles. An observer standing at a large distance from the object sees the resultant diffraction pattern, which is the sum of all the tiny waves given off by the individual electrons.

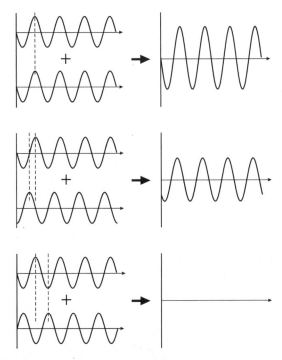

Figure 2.3. Adding two cosine waves of the same wavelength yields a third wave of the same wavelength. However, the phase and amplitude of this resultant wave, in general, will not be the same as those of the two summed waves. In fact, the relative phase of the two summed waves critically affects the appearance of the third wave. In the upper panel, the two waves to be added are perfectly in phase, and the resultant wave has an amplitude twice that of the two input waves, and the same phase. In the lower panel, the two added waves are 180° out of phase and cancel one another so that the resultant wave has zero amplitude. In the middle panel, the two added waves are out of phase by an angle somewhat less than 180°, and the resultant wave has an intermediate amplitude and phase. Two cosine waves add to give a third cosine wave. You can add this third wave to a fourth, and so on. Therefore, the sum of any number of cosine waves is another cosine.

Consider what happens when we image the diffraction pattern onto a surface. Every point on this surface is struck by a wave that is the sum of diffracted rays emanating from all points within the crystal. The different waves that strike different parts of the surface all have the same wavelength, but they differ in amplitude (how bright the light is) and in phase. The amount by which the peak of one wave is displaced from the peak of another is called the *phase difference*

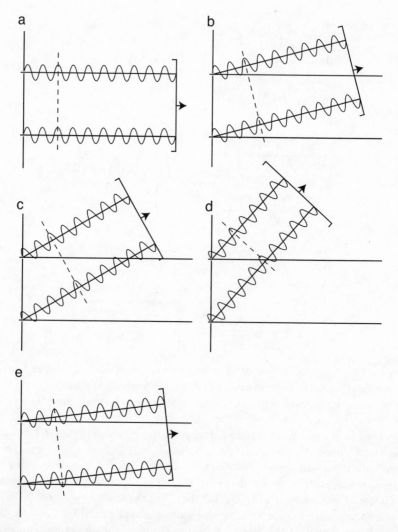

Figure 2.4. Diffraction from two adjacent holes. The two holes shown are part of a simple one-dimensional crystal, an infinite row of evenly spaced holes extending in both directions (only two holes are shown for clarity's sake). Light shines on the holes from the left, and each hole produces a spherical pattern of diffracted light rays extending off in all directions. In certain directions the waves from all the holes add up coherently, producing strong diffraction; in other directions, the waves from the different holes cancel one another, leading to no or weak diffraction. Three types of examples are shown in the figure. For each example, the dashed line allows for easy comparison of the phases of the waves scattered from the two adjacent holes. (a) The waves from adjacent holes are in phase along the direction of the incident beam and add to give strong diffraction. (b)–(d) Shown are directions for which waves from adjacent holes are shifted by an integer number of wavelengths $n\lambda$—equivalently, they

between the two waves; a shift of one wavelength λ corresponds to a phase difference of 2π. By convention the reflection hidden under the transmitted beam (the *zero-order reflection*) is assigned a phase of zero degrees. The phase of any other reflection is defined as the phase difference between that reflection and the zero-order reflection.

2.2 Diffraction from One-dimensional Crystals

It is easy to see why the diffraction pattern from a simple one-dimensional crystal is confined to spots. One-dimensional crystals have long been used in the field of optics, where they are called *gratings* or *diffraction gratings*. Imagine a grating that consists of a long line of evenly spaced holes. The grating is illuminated by a light beam, and Figure 2.4 shows light rays emanating from two neighboring holes. In general, waves emerging in a particular direction from two adjacent holes will be out of phase with one another and will not add constructively, so one sees no diffraction in these directions. However, in a few specific directions the rays from one hole will be a whole number of wavelengths out of phase with the waves from the adjacent hole.[*] Thus in these particular directions (and only in these directions), the waves from all the holes add up strongly, and a diffracted ray will be seen to emerge. As shown in Figure 2.5, the angles θ at which these rays emerge are given by

$$n\lambda = a\sin\theta \tag{2.1}$$

where $n = 1, 2, 3, \ldots$, and is the number of wavelengths by which adjacent rays differ in phase. By measuring the angle θ that the rays make with the main beam one can calculate the spacing a. This is a version of *Bragg's law* and is our first crystallographic calculation.

The row-of-holes grating gives a good idea of why we have a diffraction

---

have a phase shift of $2\pi n$. Diffraction is also strong in these directions. (e) Shown is a direction in which the waves from adjacent holes have a path length difference of $\lambda/2$; waves in this direction will cancel. In fact, the sum of waves from all the holes has a negligible value everywhere *except* in the special directions for which the path length difference is $n\lambda$.

---

[*]Being a whole number of wavelengths out of phase is the same as being in phase.

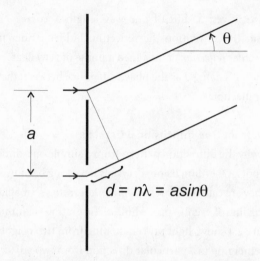

$d = n\lambda = a\sin\theta$

Figure 2.5. The geometry of Figure 2.4, shown in more detail. The two holes are separated by the crystal period a. Two diffracted rays are shown emerging from the adjacent holes, traveling in the direction specified by the angle θ. The difference in the path lengths traveled by these two waves is given by $d = a\sin\theta$. For these two diffracted rays to be in phase, this path length difference must equal a whole number of wavelengths; in other words, $d = n\lambda$, where n is an integer. The resulting equation $n\lambda = a\sin\theta$ is called *Bragg's law*. Strong diffraction will only be observed at those angles θ for which this expression is true, and thus Bragg's law explains why the diffraction pattern from a one-dimensional grating is a series of spots. We will encounter a related version of this expression for three-dimensional crystals.

Figure 2.6. A cosine grating, which is a one-dimensional crystal for which the density fluctuation takes the form of a cosine wave. This object has an ultrasimple diffraction pattern, as shown in Figure 2.7.

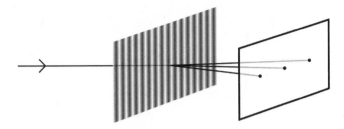

Figure 2.7. The diffraction pattern of the cosine grating in Figure 2.6. The diffraction pattern consists of only three spots, referred to as the $+1$, -1, and zeroth orders. ("Orders" is a historical usage associated with gratings; the word reflection is used in most contexts.)

pattern composed of discrete reflections, but it does not illustrate how to go from a diffraction pattern to an image. Another laser demonstration will help explain how this is done. We use a very simple specimen, the diffraction pattern of which is also very simple. Instead of a row of holes, we use a set of stripes whose profile is that of a cosine wave—that is, a densitometer trace of the specimen would show a cosine profile (Figure 2.6). The diffraction pattern of this cosine grating has only three spots. These are the zero-order reflection, which lies under the transmitted beam, and the $+1$ and -1 order reflections, which lie on either side of the transmitted beam. See Figure 2.7. (We will not explain *why* the diffraction pattern has this appearance. Simply treat it as an observation.)

We plan to build more complex patterns using this simple system, but first, we must understand what happens to the diffraction spots when we change the grating.

- If we rotate the grating, the line of three spots rotates by the same amount.
- If we stretch the grating (increase a), the spots move closer together (θ decreases). This follows from Bragg's law.
- If we shrink the grating (decrease a), the spots move farther apart (θ increases). Again, this follows from Bragg's law.
- If we make the stripes blacker, the reflections get stronger. If we make the stripes less black, the reflections get weaker. In the special case where the stripes become transparent, there is no contrast and thus no diffraction.
- If we translate the grating in Figure 2.7 parallel to itself (perpendicular to the beam), the intensity of the spots does not change. This is because the laser

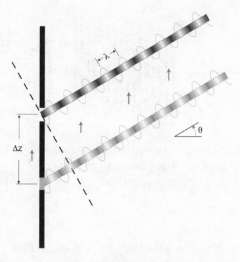

Figure 2.8. The effect of moving an object on the phase of its diffraction pattern. The object being moved is the same row-of-holes grating discussed earlier. Only a *single hole* is shown here. The original position of the hole is shown in gray; the grating is then shifted upward. Outgoing rays in the direction θ are shown for both the original and the new position of the hole. The dashed line is drawn to help us compare the positions of the wave crests before and after the hole position is shifted. It shows that the wave crests have changed position in the new ray, meaning that the phase of the diffracted ray has changed. In general, for a translation of Δz the resultant phase shift is $2\pi\Delta z \sin\theta/\lambda$. For a hole spacing of a in the one-dimensional crystal Bragg's law gives $\sin\theta = n\lambda/a$. Combining this with the previous expression gives us this result: Moving the grating a distance of Δz shifts the phase by an amount $2\pi n\Delta z/a$.

beam is much bigger than the spacing of the stripes and spans many unit cells. Translating the grating does not significantly change the number of stripes in the beam, so the amplitude of the scattered waves does not change either.

• However, when we translate the grating as described above, an invisible but very important shift is occurring: the *phase changes*.

Why does the phase change when we translate the specimen? This is a very important point, and one that is hard to explain. It is easier to first think about the row-of-holes grating. Figure 2.8 shows how the length of the path traveled by the waves changes as the grating is shifted. Look at the figure carefully. The phase shift $\Delta\alpha$ is related to the translation Δz as follows.

$$\Delta\alpha = (2\pi/\lambda)\Delta z \sin\theta$$

To gain intuition about this, imagine you are an observer standing at some point on the detector, watching the waves arrive. The wave crests would strike the detector in a regular sequence, let us say at 0, 1, 2, 3, ... time units on your clock. After the grating is moved, the wave crests will still arrive at the same intervals, but the zero time will be shifted; now the crests are arriving at 0.3, 1.3, 2.3, 3.3, ... time units (for example). Can you figure out how these shifts in the zero time are related to the phase shifts given above?

2.3 Reconstructing Images from Diffraction Patterns

We now return to the idea of building complex images using the cosine grating discussed above. In the diffraction pattern from any one-dimensional grating, you can assign each pair of spots as coming from a properly chosen cosine grating. This idea is illustrated in Figure 2.9. Here, we consider a square wave grating as an example of a simple one-dimensional crystal (this grating simply looks like a collection of evenly spaced lines). The row of spots in Figure 2.9 represents the diffraction pattern of such a square wave grating. If we consider the spots one pair at time ($\pm n$), we see that each pair by itself represents the diffraction pattern of a single cosine grating. (We have left out the zero-order reflection for simplicity.) The cosine grating with the smallest period accounts for the outer pair of spots, the one with the largest period accounts for the inner pair, and the intermediate grating accounts for the middle pair. Thus, the diffraction from the square wave grating can be approximated by the diffraction from a series of cosine gratings. Not only do the cosine gratings put spots in the right places, but other properties of the diffraction pattern are correct as well:

- The intensity of the reflection is controlled by the contrast of the grating.
- The phase of the reflection is controlled by the translational position of the grating.
- The spacing between the spots is controlled by the period of the grating.

Let us be more quantitative. We are using a series of cosine gratings to approximate a square wave grating. What are the properties of these cosine gratings? The equation describing the profile of the cosine grating that contributes the first-order spots ($n = \pm 1$) is as follows:

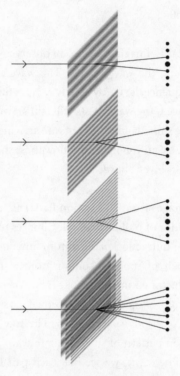

Figure 2.9. Any diffraction pattern can be represented as the sum of diffraction patterns from a series of cosine waves. The seven spots on the right represent the diffraction pattern from some arbitrary one-dimensional crystal, such as a square wave grating. The upper three panels show that individual cosine gratings can each account for one pair of spots in this diffraction pattern. Note that the periods of the cosine gratings are decreasing as we move from the top downward. The bottom panel shows the sum of all three cosine gratings, which successfully accounts for the observed diffraction pattern. Because we can account for the entire diffraction pattern of the crystal by adding together the diffraction patterns of the three cosine gratings, the sum of these three gratings must be a good representation of the crystal.

$$g(z) = F\cos\left(\frac{2\pi z}{a} - \alpha\right)$$

where g is the contrast or amplitude of the grating, F is a constant related to how dark the grating appears, a is the period of the grating (the unit cell of the crystal), and α is the phase. Remember that translating the entire grating along z changes the phase.

For the $\pm n$ pair of reflections the grating equation is

$$g_n(z) = F(n)\cos\left(\frac{2\pi nz}{a} - \alpha_n\right)$$

where the subscript n has been added to distinguish the different g's. Hence, diffraction from grating g_1 gives the $+1$ and -1 reflections; diffraction from grating g_2 gives the $+2$ and -2 reflections; diffraction from grating g_3 gives the $+3$ and -3 reflections.

And now for a very important statement:

If each pair of reflections in the crystal's diffraction pattern can be accounted for by the diffraction of a single cosine grating, then the crystal's entire diffraction pattern can be mimicked by the diffraction from the sum of many cosine gratings. If the diffraction patterns from two objects are similar, then the objects themselves must be similar; therefore, the sum of the cosine gratings must look like the crystal.

This is a profound result that introduces the field known as *Fourier analysis.* It is useful to express this result mathematically. The contrast profile for our one-dimensional crystal can be written as a sum of cosine terms. All we have to do is add up the different functions g from the individual cosine gratings. Their sum gives us the function $\rho(z)$, which represents an image of the crystal and takes the form of a sum of cosines.

$$\rho(z) = g_0 + g_1 + g_2 + g_3 + \ldots + g_n$$

$$\rho(z) = \sum_{l=0}^{n} g_l$$

$$\rho(z) = \frac{1}{a} \sum_{l=-n}^{n} F(l)\cos\left(\frac{2\pi lz}{a} - \alpha_l\right) \tag{2.2}$$

The intuitive way we developed the Fourier series results in an equation that is not exactly in standard form. For reasons of symmetry mathematicians like to include both negative and positive values of n in the series. To conform to this convention we have changed the limits of the summation in (2.2) to run from $-n$ to n. We have also added a normalization factor $(1/a)$ to conform to standard practice.

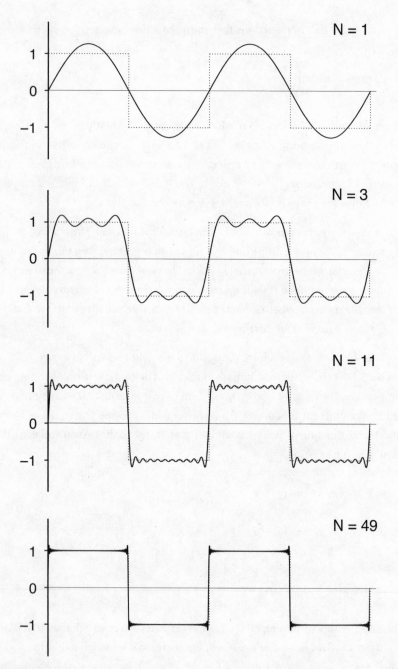

Figure 2.10. This figure shows how a grating with a square profile can be approximated by larger and larger numbers of cosine gratings. The profile of the square wave grating is shown by a dotted line. The four panels show how successively increasing N,

This result is surprising, because the object we are imaging can have any form. Why cosines, specifically? It turns out that sums of cosines can approximate *any* periodic function; Joseph Fourier demonstrated this in the early nineteenth century. Such cosine expansions are called *Fourier series* and are well known in mathematical physics. To justify our earlier assertion that we could represent a square wave grating with a series of cosine gratings, Figure 2.10 shows exactly how well a Fourier series can approximate a square wave function. This figure illustrates an important property of the Fourier series that is directly relevant to crystallography and other imaging techniques: Adding more and more cosine terms gives a better and better approximation.

Recall that in Chapter 1 we drew an analogy between microscopy and crystallography. In microscopy, the diffracted light waves are added back together by the lens, producing an image of the diffracting object. The Fourier series is the mathematical equivalent of the lens—it adds waves to produce an image. Good lenses capture more of the diffracted light than poor ones and so create a better image; similarly, the more terms (waves, or spots in the diffraction pattern) we include in the Fourier series, the more accurate our image will be.

So far in this text we have used the cosine in two very different ways: to represent cosine density fluctuations in the crystal and to generically represent the electromagnetic wave comprising the X-ray beam. Rather than using the cosine for this latter role, most books instead use the complex exponential e^{ix} (sometimes written $\exp(ix)$). This usage is based on the identity shown in Equation (2.3). The expression in (2.3) can be derived by expanding both sides in power series and showing that they are equal. This identity is far from obvious and should really be regarded as a *definition* of e^{ix}, which has no a priori meaning.

$$e^{ix} = \cos(x) + i\sin(x) \qquad\qquad (2.3)$$

Why would we use e^{ix} to denote a wave rather than the simpler cosine? One reason is that it simplifies the multiplication of two waves. The product $\cos(x) \cdot \cos(y)$ cannot be reduced, but $e^{ix} \cdot e^{iy} = e^{i(x+y)}$. Several other important identities involving e^{ix} are given below.

---

the number of cosine terms, leads to more and more accurate approximations. The top panel, with a single cosine term, is a very poor approximation; however, by the time 49 terms are included (bottom panel), the approximation is quite good.

$$e^{ix} = \cos(x) + i\sin(x)$$
$$e^{-ix} = \cos(x) - i\sin(x)$$
$$e^{ix} + e^{-ix} = 2\cos(x)$$
$$\cos(x) = \frac{e^{ix} + e^{-ix}}{2}$$

The function e^{ix} is a complex number. Recall that complex numbers take the form $A + iB$, where $i = \sqrt{-1}$. They are plotted in the complex plane as shown in Figure 2.11 and can be represented using either x-y (Cartesian) or polar coordinates. The length of the vector $C = A + iB$ is called the modulus or amplitude, written $|C|$ or C. $|C|^2 = A^2 + B^2$. $|C|^2$ can also be written $(A + iB)(A - iB)$, where $A - iB$ is called the complex conjugate of C, often written C^*. The polar coordinate expression for C is $Ce^{i\alpha}$, where C is the amplitude (vector length) and α is the phase (the angle the vector makes with the x axis). C^* is $Ce^{-i\alpha}$. This notation is particularly suitable for crystallography, because we measure amplitude and phase separately. When we measure the intensity of a reflection we are measuring $|F|^2$, where $F = Fe^{i\alpha}$. The missing phase that cannot be measured directly in the crystallographic experiment is α.

The e^{ix} notation greatly simplifies the addition of waves: Represent each wave by a complex vector, where the amplitude and phase of the vector are equal to the wave's amplitude and phase. Then add the vectors head to tail. The resultant vector has amplitude and phase equal to the amplitude and phase of the summed wave.

We can now rewrite equations (2.2) and (2.3) to represent the image of the crystal, $\rho(z)$, using the exponential notation:

$$\rho(z) = \frac{1}{a} \sum_{n=-\infty}^{\infty} F(n)e^{i\alpha_n} e^{-2\pi inz/a} \tag{2.4}$$

Note how the phase can be separated from both the amplitude and the spatial coordinate in this notation. The limits $n = \pm\infty$ are a mathematician's conceit; practically speaking, we try to measure reflections for as many values of n as possible, since adding terms adds detail to the image (illustrated in Figure 2.10). The number of terms included in the Fourier series determines the image's *resolution*. A high-resolution image is built from many terms and contains lots of detail.

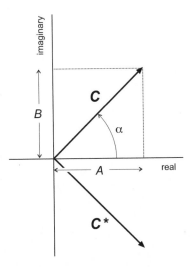

Figure 2.11. Two representations of complex numbers. In the complex plane, the real axis runs horizontally and the imaginary axis runs vertically. A complex number C is a vector in the complex plane. C can be represented by its two components, A and B, along the real and imaginary axes. In this format one would write $C = A + iB$. Alternately, C can be represented in polar form, specifying the modulus (length) of the vector C and its angle α with respect to the real axis. In this format one writes $C = |C|\exp(i\alpha) \equiv C$ $\exp(i\alpha)$. C^*, the complex conjugate of C, is also shown. C^* has polar angle $-\alpha$.

To this point we have used a square wave grating as a simple one-dimensional model for a crystal. In real crystals $\rho(z)$ represents how the electrons in the molecules that make up the crystal are distributed throughout space. Like our square wave example, the electron density function is periodic (since the crystal is made up of identical building blocks stacked together); unlike the square wave, the electron density function in crystals is three-dimensional. The next chapter will consider the diffraction from three-dimensional crystals.

2.4 Summary
- Discrete diffraction patterns result when light interacts with periodic structures.
- Periodic structures can be represented by sums of cosine terms. These sums are called Fourier series. The more terms we include in the sum, the better the quality of the resulting image.
- The diffraction from a periodic structure is equivalent to the diffraction from the summed cosine terms used to represent that object.

FURTHER READING

There are many crystallography books that discuss diffraction from various points of view.

Chapter one of *The Optical Principles of the Diffraction of X-Rays* by R. W. James (G. Bell and Sons, Ltd., London, 1954) overlaps considerably with this chapter (and reveals all the cheating that we did). James was also an explorer who reached the South Pole.

The Web site run by Kevin Cowtan, www.ysbl.york.ac.uk/~cowtan/fourier/fourier .html, provides many diffraction pattern pictures and a variety of tutorials that are relevant both to this chapter and to later ones. Your authors learned a lot from this site. Thanks, Kevin.

3

Diffraction from Three-Dimensional Crystals

3.1 The Electron Density Function in Three Dimensions

We started our discussion of diffraction using one- and two-dimensional crystals as examples because they are easy to visualize. Diffraction from three-dimensional crystals is similar to the one- and two-dimensional cases. The biggest difference is that when a three-dimensional crystal is illuminated with an X-ray beam, the reflections do not all appear at the same time. Instead, the crystal must be moved into different orientations to allow all the reflections to be observed. In the early twentieth century, the father-and-son team of William and William Lawrence Bragg developed a simple, semiquantitative model to explain diffraction from three-dimensional crystals.

In the Bragg model the crystal contains families of equally spaced parallel planes running in different directions. Examples of these families of planes are shown in Figure 3.1. The Bragg plane families are named by three integers, h, k, and l, called *Miller indices*. Within one unit cell the numbers of planes cutting the x, y, and z axes are symbolized by h (along x), k (along y), and l (along z). Thus, the 4, 1, 1 family of planes cuts the x axis four times within one unit cell; it cuts the y axis once and the z axis once. The triple of integers h, k, l is often written as a vector $\mathbf{h}$. The perpendicular distance between any two adjacent planes within a family is known as the d or Bragg spacing; d can be readily calculated from the Miller indices and the unit cell dimensions.

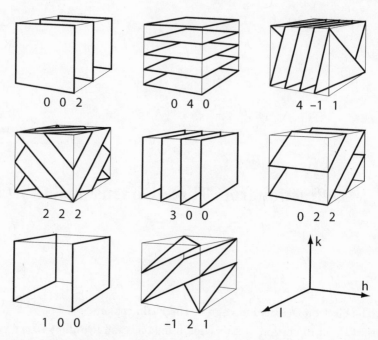

Figure 3.1. Bragg planes in a crystal unit cell. Eight different families of planes are shown. Each family h, k, l divides the x-axis into h parts, divides the y-axis into k parts, and divides the z-axis into l parts. Each of these families of planes gives rise to a distinct X-ray reflection. The values h, k, l are called Miller indices. Note that some of the indices are negative; the signs of the indices specify the direction of the tilt for that family of planes. Adapted from McPherson: *Preparation and Analysis of Protein Crystals.*

The Miller indices themselves had been developed earlier to describe the regular, external faces of crystals. It has emerged that the faces we observe in crystals are parallel to simple families of Bragg planes.

The Braggs imagined that each family of planes gave rise to a separate X-ray reflection. Each plane acts like a lightly silvered mirror, reflecting a tiny fraction of the incoming beam as shown in Figure 3.2. If (and only if) waves reflected from adjacent planes are in phase, then constructive interference occurs and strong diffraction is observed. This is strictly analogous to the one-dimensional Bragg's law case we have already discussed, when diffraction from adjacent holes adds in phase to give reflections.

As in the one-dimensional case, waves reflected from two adjacent planes are in phase only when the path length difference for the two waves equals a whole number of wavelengths. This condition leads to Bragg's law:

a.

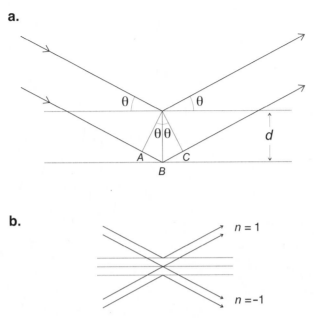

b.

Figure 3.2. Bragg's Law. (a) The two horizontal lines represent two planes belonging to a family of Bragg planes, with an interplane spacing of d. Incoming X rays enter from the upper left. Each plane acts as weak mirror and reflects a tiny fraction of each ray toward the upper right (the transmitted rays are omitted for clarity). The ray reflected from the lower plane travels a greater distance than the ray reflected from the upper plane; the extra distance is ABC = $2d \sin\theta$. When this path length difference equals a whole number of wavelengths λ, the two reflected rays are in phase and the Bragg condition occurs: $n\lambda = 2d \sin\theta$. (b) The $n = 1$ and the $n = -1$ reflection for a given set of Bragg planes. The difference between these two reflections can be understood as a difference in the orientation of the planes with respect to the X-ray beam. In one case, the rays are reflected from the "top" of the planes, and in the other they are reflected from the "bottom."

$$n\lambda = 2d \sin\theta \tag{3.1}$$

Strong diffraction is only observed for values of θ that correspond to integer values of n. In practice crystallographers only consider the $n = 1$ reflection from each family of planes, since the $n = 2$ reflection can be regarded as coming from the planes $2h, 2k, 2l$, which have a spacing $d/2$.

Representing anything as complicated as a protein crystal as a set of stacks of planes may seem like a wild oversimplification, but for the simple crystals the Braggs were studying in 1915, this model is reasonable. For example, in crystals

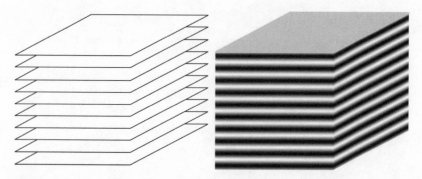

Figure 3.3. A Bragg plane family and a three-dimensional cosine density fluctuation have characteristics in common. A stack of planes and a cosine wave both having the same spacing d are drawn next to one another to illustrate this point.

of the graphite form of carbon all the atoms lie in planes located at $z = 0, 1, 2, \ldots$. These planes of atoms correspond well to the *001* family of Bragg planes. Even for more complicated crystals, the Bragg theory is very useful for understanding the geometry of diffraction. On the other hand, the parallel plane model does not lend itself to calculating images; but it is easy to connect Bragg theory with the Fourier approach we took for calculating images of one-dimensional crystals. The trick is to realize that, as shown in Figure 3.3, the Bragg plane family h, k, l resembles the three-dimensional density wave $\cos(2\pi[hx/a + ky/b + lz/c])$. The Bragg planes can be thought of as simply marking the positions of the peaks of the cosine wave. If, as in graphite, many atoms lie on or near a given set of Bragg planes, the amplitude of the cosine density wave will be large, and scattering will be strong for this reflection.

Like Bragg planes, a three-dimensional cosine density fluctuation also obeys Bragg's law. However, unlike the corresponding set of Bragg planes, the cosine density wave only gives reflections for the case $n = \pm 1$. Thus, a three-dimensional cosine density wave represents a kind of hypersimplified crystal that only gives one pair of Bragg reflections. (The $n = -1$ case occurs when the X rays come from below instead of from above, producing a symmetrically related reflection; see Figure 3.2b). Of course, real crystals will be much more complicated than a single cosine density wave, but they can be modeled by including additional density waves. Thus, just as we modeled one-dimensional functions in Chapter 2 by using sums of cosines, we can model a crystal's three-dimensional electron density distribution by using sums of cosine density waves, with each density wave contributing one pair of reflections. As we saw in Chapter 2, many cosine terms are required to accurately reproduce the fine details of the target function. To properly model the complexity of real protein crystals, tens or even hundreds of thousands of reflections are required.

The expression for the electron density function in three dimensions is closely analogous to the one-dimensional version obtained in Chapter 2. It looks quite formidable, but we are now in a position to understand it intuitively. Following is a reminder about what the variables are:

- V is the volume of the unit cell, whose edges are of length a, b, and c. For a rectangular solid, $V = abc$.
- The Miller indices h, k, l specify the direction and period of the cosine wave. They are used just like Miller indices for Bragg planes. In one unit cell there are h periods of the cosine wave along the x axis, k periods along y, and l along z.
- α is the phase of the cosine wave. It describes how far the crest of the wave is from the coordinate origin, $x = 0$, $y = 0$, $z = 0$. α is not measured directly; it is estimated by means we have not yet discussed.
- $F(h, k, l)$ is the amplitude of the cosine wave h, k, l. F is obtained from the diffraction experiment—the intensity of the h, k, l reflection is proportional to F^2.
- $\rho(x, y, z)$ is what we wish to recover, namely, the image showing where the electrons are localized in space (the electron density map). We calculate ρ at closely spaced intervals of x, y, and z to make a three-dimensional function that can be contoured and displayed on a computer screen. Calculating ρ is a big job because the summation over all reflections has to be done for each point x, y, z.

The equation in cosine form is

$$\rho(x, y, z) = \frac{1}{V} \sum_{h=-\infty}^{\infty} \sum_{k=-\infty}^{\infty} \sum_{l=-\infty}^{\infty} F(h, k, l) \cos\left(2\pi\left(\frac{hx}{a} + \frac{ky}{b} + \frac{lz}{c}\right) - \alpha\right) \quad (3.2)$$

The equation in exponential form is

$$\rho(x, y, z) = \frac{1}{V} \sum_{h=-\infty}^{\infty} \sum_{k=-\infty}^{\infty} \sum_{l=-\infty}^{\infty} F(h, k, l) \exp(i\alpha) \exp\left(-2\pi i\left(\frac{hx}{a} + \frac{ky}{b} + \frac{lz}{c}\right)\right)$$
$$(3.3)$$

Some books use vector notation to condense these equations: h, k, l = h and (x/a, y/b, z/c) = x. The latter are called fractional coordinates because they range from 0 to 1 along one unit cell edge. In vector notation, the same equation is written as

$$\rho(x) = \frac{1}{V} \sum_{-\infty}^{\infty} F(h)\exp(i\alpha)\exp(-2\pi i h \cdot x) \qquad (3.4)$$

Equations (3.2) to (3.4) are simply three different ways of writing the same equation.

In summary, the F and α for every X-ray reflection correspond to the amplitude and phase of a cosine density wave running through the crystal. By adding these density waves together we can form an image.

Figure 3.4 shows an electron density function being built of cosine fluctuations. Here we show a two-dimensional example, because it is easier to visualize than a three-dimensional one. Panels a and b show two orthogonal density fluctuations and their diffraction patterns; panel c is their sum. You can already see that with only two cosine terms, peaks are starting to appear in the image shown in Figure 3.4c. Panels d and e show the diagonal wave arising from the (1,1) reflection, but with two different phases. One wave, given by $\cos(2\pi[x/a + y/b])$, has a density maximum at the origin (in other words, its phase = 0); the other, given by $(\cos(2\pi[x/a + y/b] - \pi))$, has a minimum at the origin (phase = π radians). Panels f and g are the sums of three cosine fluctuations—the two already summed in panel c, plus the two versions of the (1,1) fluctuation. Note how the resultant images change when the phase of the (1,1) contribution is changed.

As the electron density equation explicitly states, the images of molecules that are produced using X-ray crystallography result from the addition of many cosine terms that correspond to density waves within the crystal. The example given in Figure 3.4 illustrates how critical it is to add these waves with the correct relative phases if we are to obtain an accurate image. Unfortunately, although we can measure the amplitude of each cosine term in our X-ray diffraction experiment, for practical reasons it is impossible to measure the relative phases of these terms. This is a critical technical issue and is known as the *phase problem* in crystallography. Much of the intellectual effort that has been invested in crystallography during the past century has been devoted to overcoming this problem. We discuss how protein crystallographers typically attack the phase problem in the next chapter.

The quality of the image we create depends not only on accurate phases, but

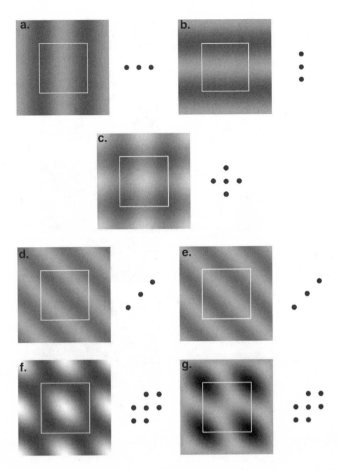

Figure 3.4. Series of two-dimensional cosine density fluctuations and their diffraction patterns. For each panel, the cosine fluctuation is shown on the left, with its diffraction pattern on the right. Panel (c) represents the sum of panels (a) and (b); (f) = (a) + (b) + (d); (g) = (a) + (b) + (e). In each panel, the white box represents one unit cell in the two-dimensional crystal. This figure is described in more detail in the text.

also on which cosine density fluctuations we incorporate in the electron density function sum. Each *hkl* term in the Fourier summation has a period *d*, equal to the Bragg spacing *d* for the *hkl* family of Bragg planes. Fourier terms with small values of *d* provide fine detail (high-resolution information) in the electron density map. Figure 3.5 illustrates the effect of resolution on the reconstruction of a single object. When high-resolution (small *d*) diffraction data are omitted from the Fourier summation, the resulting images lack detail. The same effect is seen with crystals.

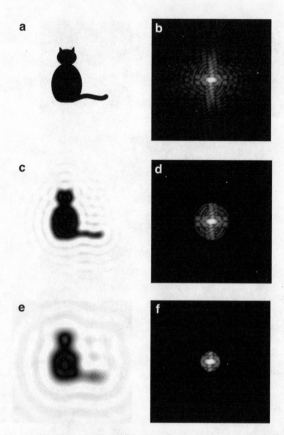

Figure 3.5. A cat and its diffraction pattern illustrate the effect of resolution on image quality. A cat is shown in (a), and its calculated diffraction pattern is in (b). (What is actually shown is the amplitude or modulus of the scattering vector—bright areas represent strong scattering, and dark places represent weak scattering). In (d), the diffraction pattern is truncated beyond a certain radius. In (c), we see the image of the cat reconstructed from this truncated diffraction pattern. (The distance from the center of the diffraction pattern corresponds to $\sin\theta/\lambda$, or resolution, so removing data beyond a particular radius is equivalent to discarding the high-resolution diffraction data.) The reconstructed image in (c) clearly does not contain the level of detail seen in (a). The resolution of the diffraction pattern is reduced even further in (f), and the corresponding image shown in (e) has even less detail. This figure and Figure 3.6 were inspired by similar figures found in Taylor and Lipson's book on optical transforms.

In the diffraction pattern shown in Figure 1.3, each spot corresponds to one reflection; the darker the spot, the more intense the reflection. Bragg's law tells us that reflections with small values of d have large values of $\sin\theta$ and so in this figure the high-resolution reflections are those spots falling near the edge of the detector. Note how the average intensity of the reflections tends to decrease as one goes from the center of the detector out to the corners. This falloff in intensity is caused in part by imperfections in the crystal that limit the resolution of the data that can be measured. Much of the art in crystallography lies in coaxing recalcitrant molecules to form well-ordered crystals from which high-resolution data can be measured.

We can formally describe how the detail of the image depends on the resolution of the diffraction pattern. The *resolution* of the map is defined as the smallest value of d for which reflections are included in the electron density sum. Thus "2 Å resolution" means that all (or most) of the terms in the summation having $d \geq 2$Å have been measured and included in the map calculation.

We can calculate d by rearranging Bragg's law.

$$\frac{1}{d} = \frac{2\sin\theta}{\lambda} \tag{3.5}$$

The quantity $2\sin\theta/\lambda$ is often used as a variable representing resolution (see, for example, Figure 3.7).

d can also be determined geometrically by calculating the spacing between adjacent Bragg planes. In the case in which the lattice contains all right angles the result is

$$\left(\frac{1}{d}\right)^2 = \left(\frac{h}{a}\right)^2 + \left(\frac{k}{b}\right)^2 + \left(\frac{l}{c}\right)^2 \tag{3.6}$$

Crystal lattices for which the unit cell axes are not orthogonal give rise to somewhat more complicated expressions for d.

Figure 3.5 represents diffraction from a single object (a cat). Most of this book, however, is concerned with diffraction from crystals. How is the diffraction pattern of a single object related to the diffraction pattern of a crystal, which contains many copies of that object? When you place multiple copies of an object next to one another, their diffraction patterns interfere with each other, just as we

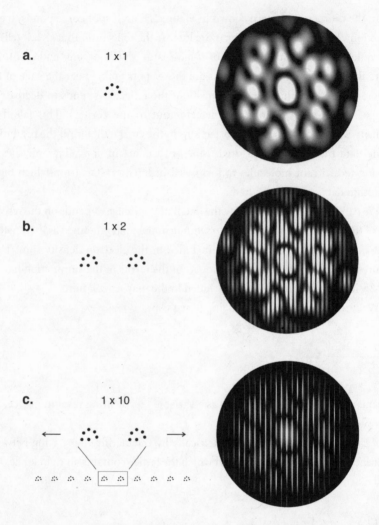

Figure 3.6. The diffraction patterns of a single object and multiple copies of that object. (a) A single object—a collection of six spots—that corresponds to a single unit cell of a crystal. Its diffraction pattern is shown on the right. Note that the diffraction pattern of a single unit cell is continuous. (b) Two side-by-side copies of the object and the diffraction pattern of this 1 × 2 array. See how placing two copies of the unit cell next to each other gives rise to interference, which causes stripes to appear in the diffraction pattern. (c) Ten copies of the unit cell assembled into a small one-dimensional crystal (a 1 × 10 array). The diffraction pattern of this one-dimensional crystal now shows pronounced sampling in the direction that corresponds to the unit cell packing. (d) A 2 × 2 array of unit cells. The diffraction pattern now begins to show sampling in both dimensions. (e) Simulation of a two-dimensional crystal with a 10 × 10 array of unit cells. The resulting diffraction

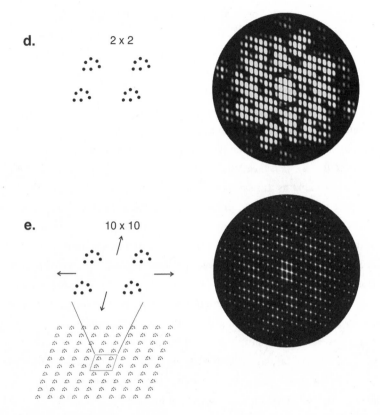

saw with the row of holes in Chapter 2. The resultant diffraction pattern will be strong in some directions and weak in others. Figure 3.6 illustrates this effect. The figure shows an object consisting of six dots (think of this array as six tiny holes in a laser diffraction grating, or as an imaginary two-dimensional molecule containing six atoms). The object is repeated in several configurations, including a two-dimensional crystalline array. The diffraction pattern of each configuration is shown at the right. You can see that the diffraction pattern of the single object is continuous; however, stacking the objects in a particular direction causes the diffraction pattern to be sampled at regular intervals in that direction. The diffraction pattern of the crystal appears very different from that of the single

---

pattern is no longer continuous, but is confined to discrete spots (almost discrete—you can still see some streaking connecting the brightest spots. In a real crystal with thousands of unit cells, even this streaking would disappear).

molecule, but if you look closely at the figure you will be able to see that they are in fact closely related. Imagine taking a piece of black paper and punching tiny holes in it at the positions of the spots in the crystal diffraction pattern. If you were to place this mask over the diffraction pattern of the single molecule, you would obtain the crystal diffraction pattern.

The diffraction pattern of a crystal is a sampled version of the diffraction pattern of the individual objects making up that crystal (the unit cells). The crystal's diffraction pattern is zero everywhere except for the spots where crystal diffraction is allowed; and at these spots the crystal's diffraction pattern equals that of the unit cell.

Bragg's law explains this sampling phenomenon. Each unit cell in the crystal produces a continuous diffraction pattern. The individual diffraction patterns from all these unit cells contribute to the net diffraction observed from the crystal. In most directions, the scattered waves from different unit cells are out of phase, and so cancel each other to give zero net scattering. Nonzero diffraction is only observed in those directions for which the diffracted waves from all the unit cells are in phase. Hence, the diffraction pattern consists of discrete spots or reflections (Bragg peaks). At the positions of these Bragg peaks, the diffracted waves from every unit cell simply add together. Therefore, at each reflection, the amplitude of the crystal's diffraction pattern equals the amplitude of the diffraction pattern of a single unit cell, multiplied by the number of unit cells in the crystal. The intensity goes up as the square of number of unit cells.

A mathematical aside: how do we represent the electron density for a non-periodic object like the cat? The diffraction patterns from crystals contain discrete reflections, and the electron density in a crystal is calculated from the sum of these reflections. As we saw in Figure 3.6, the diffraction patterns of nonperiodic objects are continuous, and so one might expect that the electron density sum would be transformed into an integral. This is exactly what happens in equation (3.7), where q, r, and s represent continuous versions of h, k, and l.

$$\rho(x,y,z) = \int\limits_{-\infty}^{\infty} \int\limits_{-\infty}^{\infty} \int\limits_{-\infty}^{\infty} F(q,r,s)\exp(-2\pi i(qx + ry + sz) + i\alpha)\, dq\,dr\,ds$$

$$(3.7)$$

This integral is called a *Fourier integral*. It allows you convert a continuous diffraction pattern to an image. This integral can be inverted (we omit the proof) to give an equation for F.

$$F(q, r, s) = \int\limits_{-\infty}^{\infty} \int\limits_{-\infty}^{\infty} \int\limits_{-\infty}^{\infty} \rho(x, y, z)\exp(2\pi i(qx + ry + sz)) \, dxdydz \qquad (3.8)$$

3.2 Calculating the Diffraction Pattern from a Known Structure

We have figured out how to make an image from the diffraction pattern. How about the other way around—a diffraction pattern from an object? This is important because after we have produced a model of a molecule in our crystal structure determination, we want to check the model to make certain it is correct. A good way to do this is to calculate the diffraction pattern of the model and to see how well it agrees with experiment.

Let us start with a crystal having one atom in the unit cell. Atoms have known distributions of electrons, and it is possible to calculate the diffraction patterns of single atoms of every type. The single-atom diffraction patterns are called atomic scattering factors and are conventionally symbolized $f(hkl)$. Values of f for all elements are available in tables; an example is shown in Figure 3.7. In general, the atomic scattering factor is normalized in terms of the scattering of one electron. Hence, $f(hkl)$ for carbon, which has six electrons, has a maximum value of six.

Atomic scattering factors are continuous, like the diffraction pattern of the cat. Suppose we have a crystal that contains a carbon atom at every lattice point and no other atoms. To calculate the diffraction pattern of this crystal, one would do what we did in Figure 3.6. Instead of the diffraction pattern of the six dots, we use the known diffraction pattern of one carbon atom. We then sample this pattern at the correct diffraction points. The resulting, very simple, equation, is given by

$$F(h, k, l) = f(h, k, l) = f\left(\frac{2\sin\theta}{\lambda}\right) \qquad (3.9)$$

The second equality comes from equation (3.5). Because the scattering is symmetric, all directions are the same; only the resolution matters.

What if the atom is not at the origin? We calculated earlier that when we shift a grating of period a by an amount z, the phase of the nth reflection is shifted by $\alpha = 2\pi nz/a$ (see Figure 2.8). This can be generalized to three dimensions. If we move an object from the point $(0, 0, 0)$ to the point (x, y, z), expressed in fractional coordinates, the phase change in the reflection hkl is $2\pi(hx + ky + lz)$. Thus, the diffraction pattern for a crystal containing one atom at a point x, y, z is:

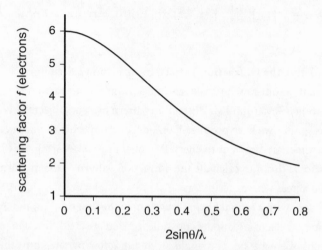

Figure 3.7. The atomic scattering factor as a function of resolution. The scattering from a single atom (if we could measure it) would give a circularly symmetric pattern on the detector. We would see a large peak of scattering at the origin, falling monotonically as one moves radially outward. The atomic scattering factor is a radial trace through this diagram. The curve shown has been calculated for carbon; other elements give curves that are very similar in appearance, differing mainly in the height of the peak at the origin.

$$F(h, k, l) = f(h, k, l)\exp(2\pi i(hx + ky + lz)) \tag{3.10}$$

Note that this expression is just like the one for an atom at the origin, except with a change in phase.

What if the crystal has many atoms in the unit cell? A crystal with many atoms in the unit cell—say N atoms—is simply the sum of N crystals, each having one atom in the unit cell. The diffraction pattern from such a crystal is calculated just like the one above, but summed over all N atoms.

$$F(h, k, l) = \sum_{j=1}^{N} f_j(h, k, l)\exp(2\pi i(hx_j + ky_j + lz_j)) \tag{3.11}$$

F is often called a *structure factor*, and equation (3.11) is known as the *structure factor equation*.

You may have noticed that we use boldface type for the structure factor F. This is to show that F is a vector, because it represents a scattered wave. We will follow this convention throughout the book. When we wish to discuss the amplitude of a vector, we will use either normal typeface or vertical bars. For example, F_P refers to a vector while both F_P and $|F_P|$ denote the amplitude of that vector. We will use both representations of the amplitude interchangeably.

How do we find the diffraction pattern starting from the electron density function, rather than an atomic model? The answer to this has already been given for the case of a single cat or other object. The answer for a crystalline array of objects is not very different and simply uses a different form of the structure factor equation. In one case we sum over individual atoms and effectively infer the electron density function from the atomic positions and the known shapes of the electron clouds of each atom. In the other case, we integrate over the whole cell using an explicitly defined electron density distribution (equation 3.8). The two forms of the equation are therefore effectively equivalent, and which form is actually used is dictated by the application.

This equation ignores the thermal motions of atoms within the crystal. Unless we're doing our structure analysis at temperatures near absolute zero (which rarely happens), each atom in the crystal will possess a significant amount of thermal energy, which causes it to vibrate. Depending on how tightly an atom is hemmed in by its neighbors, these vibrations can be substantial. This thermal motion has an interesting effect on the diffraction pattern.

The dotted arrow in the left-hand panel of Figure 3.8 represents the contribution of a single atom at point x_j to the structure factor $F(h)$. This vector is just one term in equation (3.11). If all the atoms in the crystal were perfectly still, then at any given instant the equivalent atoms in other unit cells would also lie at identical positions x_j, and each of these equivalent atoms would make identical contributions to the structure factor. This is shown in the right-hand panel of Figure 3.8 by the vector sum of many identical dotted arrows.

Because each atom undergoes thermal motion, however, each arrow oscillates around its mean position through angles $2\pi h \cdot \Delta x_j$, where Δx_j is the instantaneous displacement of the atom. The solid arrow in the left-hand panel is a snapshot of one such rotated vector—this is the contribution to the scattering factor of an atom that has strayed from its equilibrium position by Δx_j. Every one of the equivalent atoms in other unit cells is also undergoing thermal motion, but since their motions are uncorrelated, at any given instant, Δx_j will be slightly

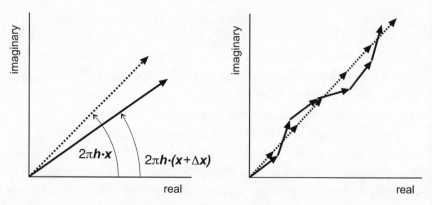

Figure 3.8. Origin of the decrease in diffracted intensity caused by thermal motions of atoms. This figure is described in the text.

different for each atom. Therefore, the actual contributions to the structure factor will resemble the vector sum of the solid arrows in the right-hand panel of the figure. The effective contribution of atom j to the structure factor is multiplied by the factor $<\cos(2\pi h\cdot\Delta x_j)>$, and so the amplitude of the summed contributions drops. (Here we have used the notation that $<z>$ is the average or expectation value of z.) As the motions Δx_j become larger, the cosine becomes smaller, and the structure factor is reduced more. Similarly, as h becomes larger, the structure factor is further reduced. This is why structure factors become weaker at high resolution.

To gain a more quantitative picture of this we need to find a way of evaluating $<\cos(2\pi h\cdot\Delta x_j)>$. If we assume a Gaussian distribution for Δx_j, we can calculate this factor.

$$<\cos(2\pi h \cdot \Delta x_j)> = \int \exp(-\Delta x^2/2u^2)\cos(2\pi h \cdot \Delta x_j)dV \qquad (3.12)$$

where $u^2 = <\Delta x^2>$, the mean square atomic displacement.

This integral is a cosine Fourier transform of the Gaussian. When evaluated, it yields another Gaussian, given by

$$<\cos(2\pi h \cdot \Delta x_j)> = \exp(-B\sin^2\theta/\lambda^2) \qquad (3.13)$$

where $B = 8\pi^2u^2$. B is called the temperature factor and is typically expressed in units of Å^2; the larger the atomic motions, the larger the value of B. The entire exponential is called the Debye-Waller factor, and it is included in the structure factor equation to account for the effects of thermal motion:

$$F(h, k, l) = \sum_{j=1}^{N} f_j(h, k, l) \exp\left(-B\frac{\sin^2\theta}{\lambda^2}\right) \exp(2\pi i(hx_j + ky_j + lz_j)) \quad (3.14)$$

Large values of B can dramatically reduce the structure factor amplitudes at high resolution. If $\lambda = 1$, for example, at 2 Å resolution $\sin^2\theta/\lambda^2 = 1/16$. If $B = 48$ Å^2 (not an unusual value for atoms in protein crystals), the Debye-Waller factor is $e^{-3} \sim 1/20$. This means that F is reduced by a factor of 20, and that F^2, which is what we measure, is decreased by a factor of 400.

High B values do not necessarily imply motion. Static disorder (i.e., a situation where equivalent atoms in different unit cells are not perfectly aligned) would have a similar effect on the diffraction. Thermal motion and other sorts of disorder are the major impediment to measuring high-resolution diffraction from most protein crystals.

3.3 Summary

- Diffraction from three-dimensional crystals can be mathematically modeled as arising from three-dimensional cosine waves (density fluctuations) running through the crystal.
- The diffraction from any individual density fluctuation can only be seen when Bragg's law is satisfied: $\lambda = 2d\sin\theta$.
- We build up an image of the crystal—the electron density function—by adding together all these density fluctuations, using a weight (F) and a phase (α) for each term. We can measure F (as F^2), but we have not yet explained how to obtain α.
- The diffraction pattern of a crystal can be calculated from its atomic model by adding together the diffraction patterns of individual atoms. The diffraction pattern of each atom is modified by a phase factor that corresponds to the atom's position in the unit cell.
- A simple model of the effects of temperature on diffraction shows that diffracted intensities are reduced by a factor $\exp\left(-B\frac{\sin^2\theta}{\lambda^2}\right)$, where B is proportional to the mean square amplitude of an atom's thermal motion. This formalism also works for other kinds of disorder.

FURTHER READING

A slender book by Ken Holmes and David Blow, *The Use of X-ray Diffraction in the Study of Protein and Nucleic Acid Structure* (Robert Krieger Publishing, Inc., New York City, 1980) covers the building up of an image exceptionally well. The book also develops the formalism of crystallography by using the mathematical tool of convolution. Convolution allows one to build a crystal from a single unit cell and a lattice. This book was originally published as a review article in *Methods of Biochemical Analysis, volume 13*. It is out of print but used copies do appear on the Web at an affordable price. A good buy.

The books by Glusker and Trueblood and by James mentioned in earlier chapters also cover the material in this chapter well. In Chapter 1 James calculates the phase difference in scattering from two objects separated by a vector distance *r* and proves in a single figure that the diffraction pattern of an object is its Fourier transform.

In the mid-1960s Taylor and Lipson published an impressive survey of optical transforms—diffraction patterns produced from tiny two-dimensional optical samples, using highly focused beams of visible light. The practical aspects of their work, while a tour-de-force of optics, are now outmoded, but the optical transforms are extremely informative, and the treatment of basic diffraction theory remains completely relevant. C. A. Taylor & H. Lipson, *Optical Transforms: Their Preparation and Application to X-ray Diffraction Problems* (Cornell University Press, Ithaca, 1965). Out of print, but available from used book sellers.

4

Phase Determination by Isomorphous Replacement

4.1 Measuring the Phases

In 1934 Bernal and Hodgkin discovered that protein crystals, when kept moist, gave rise to rich, high-resolution diffraction patterns. It became clear that these patterns contained enough information to determine the structures of protein molecules, if the phases could be measured. At that time phase determination for proteins seemed an impossible task, since estimating phases was very difficult even for simple crystals containing only tens of atoms in the unit cell. A few heroic individuals, most notably Max Perutz, continued working on the problem. In 1956 David Harker suggested a method of phase determination for protein crystals that is now called *multiple isomorphous replacement* (MIR). Perutz and others seized on this method, and within a decade, they had used it to determine the structures of several proteins, including hemoglobin and myoglobin. Since that time, multiple isomorphous replacement has been the workhorse method for phase determination in protein crystallography. The *multiple-wavelength anomalous dispersion* (MAD) method, which has recently surpassed MIR in popularity, is a derivative of MIR and is discussed later in this chapter.

To provide you with an overview of the multiple isomorphous replacement method, we list the major steps below. We will then discuss these steps one at a time.

1. Collect X-ray data $F_P^2(\boldsymbol{h})$ from your native crystal.
2. Form a heavy atom derivative of this crystal.
3. Collect data $F_{PH}^2(\boldsymbol{h})$ from the derivative.
4. Use the relation $F_H^2(\boldsymbol{h}) \approx (F_{PH}(\boldsymbol{h}) - F_P(\boldsymbol{h}))^2$ to estimate the heavy atom contribution to diffraction.
5. Use the estimated heavy atom contribution $F_H^2(\boldsymbol{h})$ to find the location of the heavy atoms in the unit cell.
6. Repeat steps 2–5 for another derivative.
7. Use the phase circle construction to calculate the missing phase for each reflection.

Collecting data means to measure the value of $F_P^2(\boldsymbol{h})$ for as many different reflections as possible (we are ignoring the nuts and bolts of data collection in this discussion). Recall that the resolution of your data is defined by the smallest value of d (largest $\sin\theta$) for which you can measure intensities for all (or most) of the reflections. The intensities that we measure in the diffraction experiment give us the square of the structure factor amplitude, so we merely need to take the square root to obtain the length of the structure factor vectors. Our first step is to collect a complete data set from a *native* crystal. By native, we mean that this is a crystal of the protein of interest, without any heavy atoms bound to it. Data collected from a native crystal are called native data.

Next we need to convert a native crystal into a heavy atom derivative. In heavy atom derivatives a high-atomic-number atom is bound to specific sites on all the protein molecules in the crystal, as shown in Figure 4.1. In the ideal case the addition of the heavy atom has no effect on the crystal lattice or the protein conformation. This, in fact, is what "isomorphous" means—if the addition of the heavy atom does not alter the protein's packing or conformation, then the native and derivative crystals are said to be isomorphous. (Obtaining perfectly iso-morphous derivatives can be a tricky business. One of the main sources of error in the MIR method is the lack of perfect isomorphism.) Derivatives are usually made by soaking native crystals in buffer solutions containing heavy atom com-pounds. Protein crystals have solvent channels through which the heavy atoms can diffuse, allowing them to bind to sites on the protein surface. Crystallogra-phers have developed a large arsenal of different heavy atom reagents that can bind to different protein side chains. Mercurial reagents tend to bind to cysteine side chains, for example, and histidine side chains can coordinate platinum, gold,

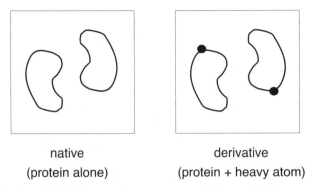

native	derivative
(protein alone)	(protein + heavy atom)

Figure 4.1. Schematic representing the preparation of heavy atom derivatives. At left is shown a unit cell of a protein crystal, containing a pair of symmetry-related protein molecules—the native crystal. At right is shown the heavy atom derivative. In the derivative, the protein's structure remains unaltered, but a heavy atom has been bound to a specific site on every protein molecule.

and similar metals, while glutamate and aspartate side chains can coordinate positively charged lanthanide ions.

Why do we want to add heavy atoms? Heavy atoms have lots of electrons, and electrons scatter X rays; even one or two heavy atoms per protein can significantly change the scattering from a crystal. By adding heavy atoms, we are perturbing the scattering in a controlled way. We will learn shortly how we can exploit these perturbations to estimate the phases.

How heavy is "heavy"? The answer depends on many experimental variables, but in general, iodine (53 electrons) is about the lightest useful element. Some of the more commonly used heavy atom reagents contain platinum, mercury, or lead (78, 80, and 82 electrons, respectively). Mercury acetate and uranyl nitrate are two examples of popular heavy atom compounds.

The addition of the heavy atom causes small changes in the amplitude and phase of each reflection. The vector diagram in Figure 4.2 shows the vector F_P representing the phase and amplitude of some reflection, as measured from a native crystal, and also F_{PH}, which represents the same reflection measured from the crystal of the heavy atom derivative. The scattering of the derivative is made up of the scattering from the protein F_P plus the additional scattering contributed by the heavy atoms, F_H (F_P and F_H sum to give F_{PH}). Note that the phase α of the heavy atom scattering is *random* with respect to the phase of the protein component. That is, the angle between vectors F_P and F_H can have any value. The

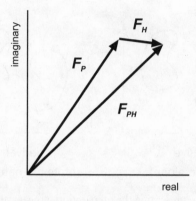

Figure 4.2. The vector triangle formed by the scattering from the protein F_P, the heavy atom F_H, and the derivatized protein F_{PH}. One can draw a diagram like this for every reflection.

amount by which the amplitude of a given reflection changes upon adding the heavy atom, $|F_{PH} - F_P|$, is known as the *isomorphous difference.*

F_H is the scattering that our crystal would produce if it contained only the heavy atoms (magically suspended in space, with no protein molecules to support them). Obviously, we cannot measure F_H or even $|F_H|$. The only experimental data that we have are $|F_P|$ and $|F_{PH}|$, but we can crudely approximate $|F_H|$ by the isomorphous difference, that is, $|F_H| \approx ||F_{PH}| - |F_P||$. Note that this expression is only truly correct when F_P and F_H are collinear; the remainder of the time, it is an approximation, and depending on the phase α_H, it can be a pretty poor one. However, it is what we have to work with.

Why are we so interested in $|F_H|$? Because we can use $|F_H|$ to deduce the positions of the heavy atoms. Many methods are available for finding the locations of the heavy atoms, but they all require some estimate for $|F_H|$. We will first discuss a method that is not the most common, but which is easy to understand and gives you the flavor. The process is a simple trial-and-error one in which we systematically test each sample point in the unit cell to see whether it is a potential site for the heavy atom. We divide the unit cell into a fine grid—say 1 Å on edge. We place an atom at grid point number one, and then use the symmetry of the crystal to place atoms at other, symmetry-related grid points in the unit cell. This process might yield, for example, 2, 4, or 8 atoms in total. We then calculate the diffraction pattern $F_{H\,calc}$ from this set of atoms using the structure factor equation (see equation 4.1).

If we have guessed the wrong location for the heavy atom, $|F_{H\,calc}|$ will not be similar to our estimate of $|F_H|$, but if we have guessed the correct one, the calculated and observed values will be similar. We repeat this calculation for every grid point, and the point for which $|F_{H\,calc}|$ and $|F_H|$ are most similar is taken to be the correct position of the heavy atom.

A more commonly used method for finding heavy atoms involves the use of the *Patterson function*. This function is convenient to use, because it can be calculated from the structure factors' experimentally measured amplitudes and it requires no knowledge of their phases. The Patterson function represents a three-dimensional map of the vectors between atoms, from which the actual atomic positions can be inferred. The Patterson function is discussed in Chapter 5.

Having determined the locations of the heavy atoms, we can apply the structure factor equation to calculate the amplitude and phase of the heavy atom scattering:

$$F_{H\,calc}(\boldsymbol{h}) = \sum_j f_{H,j}(\boldsymbol{h})e^{2\pi i \boldsymbol{h}\cdot x_j} \tag{4.1}$$

(The initial values of the heavy atom coordinates $\boldsymbol{x}_j$ are usually in error, and in general, they are refined by processes like those described in Chapter 7 before proceeding to the next step.)

Now we can determine the phase—almost. We know three things: the amplitudes $|F_P|$ and $|F_{PH}|$ and the vector $\boldsymbol{F}_H$. We can assemble these items in a diagram known as the *phase circle construction*, shown in Figure 4.3a. You prepare the phase diagram as follows: Draw a circle with radius $|F_P|$, centered at the origin. We know that the vector $\boldsymbol{F}_P$ has its tip somewhere on this circle but we do not know where (because we only know its amplitude and not its phase). Also, from the origin draw the vector $-\boldsymbol{F}_H$, whose phase and amplitude we have calculated from the heavy atom positions. The terminus of this vector lies at point P. Next draw a circle of radius $|F_{PH}|$ centered at P. Again, the vector $\boldsymbol{F}_{PH}$ has its tip somewhere on this circle, but we do not know where. The points along this circle represent all possible values of the vector $-\boldsymbol{F}_H + \boldsymbol{F}_{PH}$ (or $\boldsymbol{F}_{PH} - \boldsymbol{F}_H$), corresponding to all the possible phases of $\boldsymbol{F}_{PH}$.

There are only two points at which all the information is consistent. Remember that the vector $\boldsymbol{F}_P = \boldsymbol{F}_{PH} - \boldsymbol{F}_H$ (Figure 4.2). Because $\boldsymbol{F}_P$ must lie on the circle centered at the origin, and $\boldsymbol{F}_{PH} - \boldsymbol{F}_H$ must lie on the circle centered at P, only the

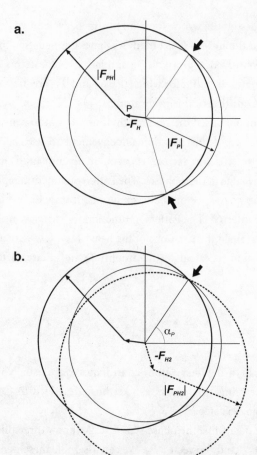

Figure 4.3. The phase circle construction for a single derivative (a) and for two derivatives (b). There is one diagram like this for each reflection. The construction of these phase circles is explained in the text. Only at the point marked by the heavy arrow in (b) are the vector triangles from both derivatives consistent. In the real world, because of experimental error, there is usually no single point that is consistent with all the data, and elaborate algorithms are used to find the best phase.

points where these two circles intersect will satisfy all the constraints on F_H, F_P, and F_{PH}. Thus, F_P must terminate at either of the two points indicated by heavy arrows in Figure 4.3a, and α_P can only take on the two possible values shown. But which one is correct? This diagram cannot tell us. We need a second heavy atom derivative to resolve the ambiguity.

Two heavy atom derivatives can determine the phase unambiguously. Figure

4.3b shows a phase circle diagram that incorporates information from a second, independent heavy atom derivative. In the second derivative, the heavy atoms occupy different positions than they do in the first, so its heavy atom scattering vector F_{H2} is different from F_H. This leads to a different set of phase circles for the second derivative. Because the protein scattering F_P is the same for the native and both derivative crystals, all the phase circles must intersect at one point, which is shown by the single heavy arrow. Only this point is consistent with the information from both derivatives, so the value of α_P corresponding to the heavy arrow must be the correct one.

In the real world phases determined in this way are subject to large errors. Because errors exist in F_H, F_P, and F_{PH}, there is usually no single point where all the circles meet. To improve the situation, more than two derivatives are often used. However, the problem of choosing the correct phase when the circles do not intersect cleanly remains a thorny one, and elaborate statistical procedures have been devised to make the best choice and estimate the associated error. The *figure of merit* for a reflection is an estimate of the reliability of the phase for that reflection. Often denoted m, the figure of merit ranges from 0 to 1 and is approximately equal to the cosine of the phase error for that reflection. The mean figure of merit is the average of m over all reflections. A mean figure of merit of 0.7, for example, means that the average phase is in error by 45°.

Once the phases have been estimated, we can calculate the electron density function $\rho(x, y, z)$ using the expression given in Chapter 3 (equations 3.2–4). Blow and Crick showed that the use of m-weighted coefficients in this summation can reduce the mean squared error in ρ. Thus, equation (3.4) is modified so that $mF(h)$ replaces the original $F(h)$. This function is then contoured and displayed on a computer, and a molecular model is fit as described in Section 4.3.

4.2 MAD Phasing

A new phasing technique called MAD has succeeded multiple isomorphous replacement as the dominant method. MAD (<u>m</u>ultiple-wavelength <u>a</u>nomalous <u>d</u>ispersion) is a variant of isomorphous replacement. In multiple isomorphous replacement the scattering of a heavy atom at a known position acts as a reference beacon—once we deduce the heavy atom substructure, we can calculate its scattering and use that information to phase the remainder of the diffraction pattern. In MAD, we also use a small number of special atoms as reference beacons. We alter the scattering of these atoms by changing the wavelength λ of the X rays.

This change in scattering allows phases to be determined. MIR and MAD are highly analogous. In both cases, we perturb the structure in a way that alters the diffraction. For MIR, the perturbation is the addition of a heavy atom; for MAD, the perturbation is a change in wavelength. For both methods, we characterize the nature of the perturbation, calculate how the perturbation affects the diffraction, and use this information to phase the diffraction pattern.

How does changing the wavelength of the X rays change the diffraction properties of an atom? First, you should remember that changing the wavelength of X rays also changes their energy. Next, we must explain that up to this point, when we calculated how atoms scatter X rays, we assumed that the electrons were "free"—in other words, the electron's scattering properties were not affected by the fact that it is part of a larger atomic structure. For free electrons, scattering does not change drastically when the wavelength is changed. When the energy of the X rays is very close to the energy of a transition of an electron from one shell to another, however, the assumption of free electrons is no longer valid. Under these conditions, the electron exhibits resonant behavior that affects its scattering properties. Resonant effects such as this are often seen in everyday life, for example, when a car starts to vibrate at a certain speed.

4.2.1 How Is Anomalous Scattering Different from Normal Scattering?

Fortunately, "anomalous" scattering is a misnomer—we actually understand the process reasonably well. When λ approaches the wavelength where an electronic transition occurs (also known as an absorption edge), both the *strength* of the scattering and the *phase* of the scattered radiation change. We express these changes using a generalized form of the atomic scattering factor f:

$$f_{\text{anom}} = f_o + f' + if'' \qquad (4.2)$$

f_o is the scattering from the nonresonant atom, that is, the scattering that one would observe far from the absorption edge. f_o is independent of wavelength. Near the absorption edge, the strength of the atom's scattering begins to change quite rapidly with wavelength. This effect is described by f', the wavelength-dependent change in the normal scattering. Finally, a new scattering term if'' appears near the absorption edge; scattering due to this term is 90° out of phase with the normal scattering and is wavelength dependent. We will not attempt to derive why any of these changes in the scattering factor occur near the absorption

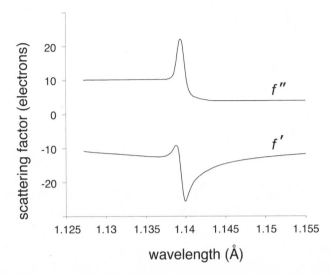

Figure 4.4. The wavelength dependence of f' and f'', the real and imaginary contributions to anomalous scattering, for an osmium derivative of a macromolecule. These data were obtained by one of the authors, using tunable X rays at a synchrotron radiation source. The LIII edge for osmium lies at 1.14 Å. Note how sharply f' and f'' change with minute changes in wavelength near the edge.

edge, but they will prove to be very important in exploiting the anomalous scattering phenomenon for the purpose of phasing.

Figure 4.4 shows the change in f' and f'' with wavelength (or energy) for a representative element, osmium. Note that the strength of the scattering effect is expressed in units of electrons. At the peak wavelengths, the magnitude of the anomalous scattering effect for osmium is about 25 electrons. This means that for every osmium atom bound to the protein, altering the wavelength can perturb the scattering by an amount that is equivalent to adding a heavy atom of atomic number 25. This is not a very heavy atom, in particular when compared with the heavy atoms typically used in multiple isomorphous replacement. However, the small size of the perturbation is compensated by the fact that the structure of the crystal remains unchanged when we alter the wavelength, and therefore, data measured at different wavelengths can be thought of as deriving from perfectly isomorphous derivatives.

Unfortunately, none of the elements normally found in proteins—C, H, N, O, S—exhibit strong anomalous scattering effects (sulfur is a marginal anomalous scatterer). Sometimes anomalous scatterers can be introduced into the crystal by soaking, just as heavy atoms are introduced for the MIR method. In other cases, the protein may naturally contain a prosthetic group carrying an anomalous scatterer (iron, for example). The most general approach to anomalous scattering studies utilizes selenium. Although selenium is not found in proteins naturally, the selenium analog of methionine, called selenomethionine, can be introduced into the growth medium of bacteria and thus into recombinant proteins. Although the anomalous scattering effect associated with selenium is not as strong as that with osmium, it is still significant and has been very successfully exploited for phasing protein structures.

As the name suggests, MAD experiments are carried out using more than one wavelength, for reasons we illustrate below. At least two different wavelengths are required, and they are typically chosen to maximize f'' and minimize f', respectively. Almost always, a third wavelength is chosen at an energy remote from the absorption edge.

We have known about the anomalous scattering effect for a long time, but to exploit it for phasing, we must be able to adjust the wavelength of the X rays, just as one changes the wavelength in a spectrophotometer. Laboratory X-ray sources have a fixed wavelength, and so it was only when synchrotrons arrived, with their very intense, tunable beams, that this technique came of age. It was pioneered by Wayne Hendrickson in the 1980s.

A key feature of MAD is the breakdown of what is known as *Friedel's law*. Friedel's law says that $F(h,k,l) = F(-h, -k, -l)$. Recall from Chapter 3 that $F(h,k,l)$ and $F(-h, -k, -l)$ can be considered to be rays reflected from opposite faces of the same set of Bragg planes. Such pairs of reflections are said to be centrosymmetrically related, and both members of the pair normally have the same intensity. When anomalous scatterers are present this no longer holds. The small differences between $F(\boldsymbol{h})$ and $F(-\boldsymbol{h})$ are key to phase determination.

The algebra required to explain MAD phase determination is beyond the scope of this text, but Figures 4.5 and 4.6 give the basic idea. These figures are analogous to Figures 4.2 and 4.3, which pertain to multiple isomorphous replacement. Figure 4.5 describes the scattering that is observed from a crystal containing a few anomalous scattering atoms and many nonanomalously scattering atoms—for example, a protein containing a few selenomethionine residues. In Figure 4.5a, $\boldsymbol{F_N}$ is the scattering for all the nonselenium atoms. f_{ao} is the normal scattering from the selenium atoms—the scattering that one would see at wavelengths far from the absorption edge. f'_a is the change in scattering due to the f'

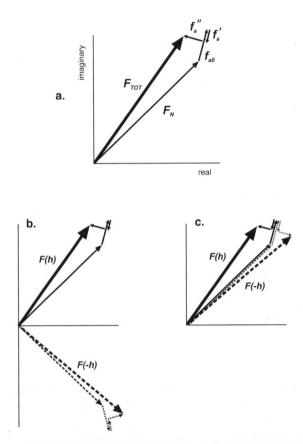

Figure 4.5. (a) The vector triangle associated with the anomalous scattering effect. As explained in the text, F_N is the scattering from the nonanomalously scattering atoms in the crystal. The scattering from the anomalous atoms is given by the resultant of three vectors: The wavelength-independent f_{a0} vector plus the wavelength-dependent f'_a and f''_a vectors. The total scattering measured from the crystal, F_{TOT}, is the vector sum of all of these scattering vectors. (b) The relationship between $F(h)$ and $F(-h)$ when anomalous scatterers are present. For the normal scattering components, the phase of any reflection h is the negative of the phase of the $-h$ reflection; thus the normal scattering components of $F(h)$ and $F(-h)$ are mirrored in the x axis. However, the imaginary scattering term if'' is always 90° out of phase with the normal scattering terms. This means that the imaginary scattering components of h and $-h$ are *not* mirrored in the x axis. As a result, the length of the resultant vectors $F(h)$ and $F(-h)$ will not be equal. This is known as the breakdown of Friedel's law and is seen most clearly in (c), where the components of $F(-h)$ have been flipped across the x axis to allow direct comparison with $F(h)$.

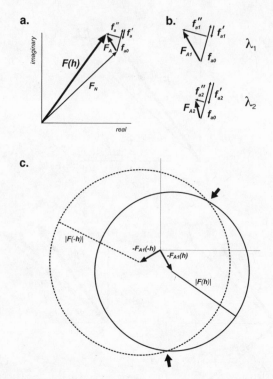

Figure 4.6. The construction of phase circles for MAD phasing is analogous to that for multiple isomorphous replacement. (a) To simplify the phase diagram, we represent the vector sum of the different components of the anomalous scattering with a single vector, $F_A = f_{a0} + f'_a + f''_a$. (b) Because f'_a and f''_a are wavelength dependent, both the phase and amplitude of F_A change when the X-ray wavelength is changed from λ_1 to λ_2. (c) Construction of the phase circle diagram for wavelength λ_1. Starting from the origin, we draw the vector $-F_A(h)$. Centered on the tip of this vector, we draw a circle of radius $|F(h)|$. This circle represents all possible values of the vector $F(h) - F_A(h)$, corresponding to the different possible phases for $F(h)$. The procedure is then repeated to generate a second circle representing all possible values for the vector $F(-h) - F_A(-h)$. These circles intersect at two points, indicated by the two heavy arrows. Just as in the case of a single isomorphous derivative, we cannot distinguish which of the two possibilities corresponds to the true phase α_N; data from another wavelength are required to resolve the ambiguity, as shown in Figure 4.7.

term in the atomic scattering factor, and f''_a is the change in scattering due to f''. Note that f''_a is perpendicular to both f_{a0} and f'_a (because it is 90° out of phase with the normal scattering). What we actually observe is the result of all these scattering vectors, namely, the total scattering $F = F_N + f_{a0} + f'_a + f''_a$.

Figure 4.5b shows how $F(h)$ and its centrosymmetrically related reflection

$F(-h)$ are affected when anomalous scattering is present. For *normal* scattering, the phase of any reflection $\alpha(h) = -\alpha(-h)$. Thus, $F_N(h)$ and $F_N(-h)$ form a pair of vectors reflected through the x axis; the same is true for both f_{a0} and f'_a. However, the f'' term is always 90° out of phase with the normal scattering term, which means that $f''_a(h)$ and $f''_a(-h)$ do not form such a mirrored pair of vectors. The result is that $|F(h)|$ and $|F(-h)|$ are no longer equal when a significant anomalous scattering component is present. This can be seen clearly in Figure 4.5c, where $F(-h)$ has been reflected across the x axis to allow for an easy comparison with $F(h)$.

This diagram contains information similar to that in a multiple isomorphous replacement diagram. To exploit this information for phasing, we start by measuring the lengths of the vectors $F(h)$ and $F(-h)$ (in other words, we collect a data set at a given wavelength). We then deduce the positions of the anomalous scatterers. There are usually only a few such scatterers, relative to the total number of atoms in the protein. To find their positions we use methods analogous to those used in multiple isomorphous replacement to find heavy atoms, except we use the anomalous differences $|F(h) - F(-h)|$ instead of the isomorphous differences.

Once we have located the anomalous scatterers, we can calculate *the lengths and phases* of f_{a0}, f'_a, and f''_a. This allows us to construct a phase circle diagram (Figure 4.6). Note that the anomalous differences measured at a single wavelength do not allow us to unambiguously determine the phases—we are left with a twofold ambiguity, analogous to the isomorphous replacement situation in which we have a single derivative. We need more information. Because f' and f'' change with wavelength, the phase ambiguity can be overcome by making another measurement of the anomalous differences at a different wavelength, as shown in Figure 4.7.

A careful examination of Figure 4.6 reveals the essential similarity of the MAD and MIR methods. If we think of the anomalous scatterers as being analogous to heavy atoms, then F_N in a MAD experiment is equivalent to the native vector F_P in the MIR experiment, and $F(h)$ and $F(-h)$ are equivalent to F_{PH} values for two different derivatives. The phase circle construction drawn in Figure 4.6c is therefore similar to the two-derivative MIR experiment shown in Figure 4.3b. There is a key difference, however—*we do not actually measure the value of F_N in the MAD experiment*. Therefore, $F(h)$ and $F(-h)$ data for a single wavelength are not sufficient to resolve the phase ambiguity in a MAD experiment—data from at least one additional wavelength are also required.

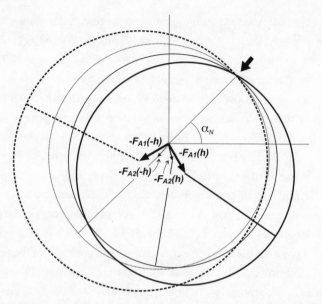

Figure 4.7. MAD data from two different wavelengths provide a unique solution for the phase circle construction. A phase circle diagram is shown that incorporates information from both wavelengths λ_1 and λ_2 shown in Figure 4.6. All four circles are seen to intersect in a single point (marked with the heavy arrow), corresponding to the correct phase.

It is possible to combine the anomalous scattering and isomorphous replacement methods. Many of the heavy atoms commonly used for isomorphous replacement have strong anomalous scattering signals, which means that $F_{PH}(\boldsymbol{h})$ will differ significantly from $F_{PH}(-\boldsymbol{h})$. Hence, a single derivative with anomalous scattering is equivalent to two different derivatives. We call this method of phasing <u>s</u>ingle <u>i</u>somorphous <u>r</u>eplacement with <u>a</u>nomalous <u>s</u>cattering, or SIRAS.

4.3 Fitting Models to Experimental Electron Density Maps

The phasing techniques described in this chapter are intended to produce an electron density map. This map represents a direct image of the electrons within the molecule, but it is not the most useful representation of a protein. Chemists and biologists prefer to view molecules in the form of atomic models, in which the atomic positions are linked by sticks representing chemical bonds. Once phase estimates have been obtained and the map calculated, therefore, the crystallographer's next task is to interpret the map by fitting an atomic model.

Fitting the map is typically one of the most labor-intensive steps in a crystal structure determination. Conceptually, the problem is straightforward. We must determine which portions of the map correspond to the different amino acids of the protein and place the atoms of these residues accordingly (refer to Figure 1.4 for an example of a well-fit electron density map). We are aided in this process by two things: We know the amino acid sequence of the protein, and we know a great deal about protein stereochemistry (bond lengths, angles, and so on). This knowledge is sufficient to construct a stereochemically reasonable model of the protein. What remains to be determined are the molecule's torsion angles. These angles determine the path followed by the protein's backbone and the conformations of its side chains. The fitting procedure is equivalent to adjusting the torsion angles so as to "thread" the protein chain through the electron density. In principle, one could build the model by positioning the first amino acid of the chain into its proper place in the electron density map and then adjusting the torsion angles of each subsequent residue in turn to position the atoms within electron density.

In practice, the procedure is complicated by many factors. For example, side chains are the landmarks that allow us to map the sequence onto the electron density map. However, many crystals diffract to only modest resolution, yielding maps that lack detail and making it difficult to distinguish the side chains of different amino acids. Also, estimated phases frequently contain significant errors, leading to noisy electron density maps. The convoluted paths followed by protein chains are difficult to follow in noisy maps. This makes tracing the path of the chain like finding one's way out of a maze; dead ends are commonly encountered, so a trial-and-error approach is required. The situation is complicated by the close packing of protein molecules in crystals—it is often difficult to decide where one molecule leaves off and its neighbor begins.

Fitting a protein chain to an electron density map requires complex judgments that are difficult to reduce to simple formulas. Consequently, it has been challenging to automate the process, and reliable fitting programs have only been developed relatively recently. Not surprisingly, these programs perform better with high-quality maps than with poor maps. Maps at resolutions of 2 Å or better calculated with accurate phases can frequently be completely fit without manual intervention by a crystallographer. Maps at lower resolution and/or with high noise levels, however, still require substantial hands-on effort to interpret.

Since the 1970s, fitting has been accomplished using software that allows the user to build and manipulate stick figures of molecules, overlaid on displays of the electron density map. Before the development of computer graphics, many protein models were fit by using an ingenious optical comparator device known as a Richards box (named after its inventor, the protein chemist Frederic Richards of Yale University). Wire models were built by hand out of thin metal rods. Each chemical bond was represented by a small rod several centimeters long; at this scale, the finished molecular model could span several meters. Sections through the electron density were plotted on Plexiglas plates, which were stacked to provide a three-dimensional view of the map. A semitransparent mirror was then placed between the model and the map, and the lighting was arranged so that the model appeared superimposed on the map. The model was adjusted until it fit the map well, and coordinates were then measured from the model with rulers. The Richards box was also affectionately called *Fred's Folly*.

4.4 Summary

- In multiple isomorphous replacement, protein crystals are derivatized by specific binding of heavy atoms. The positions of these heavy atoms in the unit cell can be determined by using the isomorphous differences $|F_{PH} - F_P|$. The heavy atom structure factors F_H can then be calculated. The heavy atoms serve as reference points that allow us to find the protein phases.

- MAD phasing relies on the presence in the unit cell of an anomalous scatterer, such as selenium, that plays a role analogous to a heavy atom. When data are collected at wavelengths where the anomalous effect is strongest, the change in scattering by the selenium echoes the change in scattering upon addition of a heavy atom. Phases can be extracted by methods similar to those used in multiple isomorphous replacement.

- Once the electron density map has been calculated, the map must be interpreted by fitting an atomic model to it. Building an atomic model that matches the protein sequence and contains appropriate bond lengths and angles is straightforward, but adjusting the model's torsion angles so that the conformation of the protein matches the electron density is a more complex task.

FURTHER READING

Phasing by isomorphous replacement and by anomalous scattering are covered in depth by Blundell and Johnson in *Protein Crystallography* (Academic Press, New York,

1976). This text, which has long served as the macromolecular crystallographer's bible, also covers the figure of merit with rigor and is a good reference for the chemistry of heavy atom ligands. This book is out of print, and used copies have sold for over $700. Keep your old texts!

The more contemporary book by Jan Drenth, *Principles of Protein X-ray Crystallography*, 2nd edition (Springer-Verlag, New York, 1999), has a concise style and presents a good treatment of MAD, as well as of many other topics. It is strong on analysis of errors (and contains in-your-face math!).

Wayne Hendrickson, the architect of MAD, provides an insightful and compact introduction in his article "Determination of macromolecular structures from anomalous diffraction of synchrotron radiation," *Science* 1991; 254(5028):51–58. Explains with great clarity how one best combines anomalous scattering measured at several wavelengths.

The Patterson Function

5.1 Definition of the Patterson Function

Remember that using MIR or MAD to solve the phase problem requires that we determine the positions of the relevant heavy atoms or anomalous scatterers within the unit cell. The Patterson function is a powerful tool for finding these atoms. The function was defined in the early 1930s by A. L. Patterson, who wondered what information he could gain if he computed a Fourier synthesis from a diffraction pattern for which he didn't know the phases. He arrived at the function

$$P(\boldsymbol{u}) = \frac{1}{V} \sum_{\boldsymbol{h}} F^2(\boldsymbol{h}) \exp(-2\pi i \boldsymbol{h} \cdot \boldsymbol{u}) \qquad (5.1)$$

in which he set all the phases to zero.* $P(\boldsymbol{u})$—now known as the Patterson function—resembles the electron density function $\rho(\boldsymbol{x})$. It is defined in the same unit cell as the electron density function, but the amplitude of the coefficients is chosen to be $F^2(\boldsymbol{h})$ instead of $F(\boldsymbol{h})$. To avoid confusing the Patterson and electron density functions, it is customary to use a variable other than $\boldsymbol{x}$ to describe the space of P. We have chosen $\boldsymbol{u}$.

Unlike the electron density function, which cannot be calculated until we know the phases, the Patterson function can always be calculated, because the

---

*The scalar or dot product of two vectors $\boldsymbol{u}$ and $\boldsymbol{v}$ is typically written $\boldsymbol{u} \cdot \boldsymbol{v}$; however, it is sometimes written as simply $\boldsymbol{uv}$. We use the two conventions interchangeably.

phases are set to zero—only the experimental intensity data are required. However, does the Patterson function have any useful physical meaning? It turns out that it does.

In the electron density function peaks correspond to atoms. In the Patterson function peaks correspond not to the positions of the atoms themselves, but to the *vectors between atoms*. Thus, if the Patterson function has a peak at vector position **u**, it means that there is a pair of atoms (or many pairs) that are separated by the same vector distance **u**. To help demonstrate this, we can use the structure factor equation to obtain an expression for F^2. Recall from Chapter 2 that the squared modulus of any complex number **C** is given by the product of **C** with its complex conjugate, that is, $|C|^2 = C \cdot C^*$, where **C*** denotes the complex conjugate of **C**. Applying this to the structure factor equation, we obtain the following:

$$F^2 = F \cdot F^*$$

$$F(h) = \sum_{\substack{\text{atoms} \\ j=1}}^{N} f_j \exp(2\pi ih \cdot x_j)$$

$$F^*(h) = \sum_{\substack{\text{atoms} \\ k=1}}^{N} f_k \exp(-2\pi ih \cdot x_k)$$

$$F(h) \cdot F^*(h) = \sum_{\substack{\text{atoms} \\ j=1}}^{N} f_j \exp(2\pi ih \cdot x_j) \times \sum_{\substack{\text{atoms} \\ k=1}}^{N} f_k \exp(-2\pi ih \cdot x_k)$$

$$F^2 = \sum_j \sum_k f_j f_k \exp(2\pi ih \cdot [x_j - x_k]) \tag{5.2}$$

We find that F^2 depends on the set of interatomic vectors $x_j - x_k$. In contrast, the structure factor **F** depends on the set of atomic positions, x_j:

$$F(h) = \sum_{\substack{\text{atoms} \\ j=1}}^{N} f_j \exp(2\pi ih \cdot x_j) \tag{5.3}*$$

---

*Equation (5.3) is equivalent to equation (3.11) (but written with a slightly different notation).

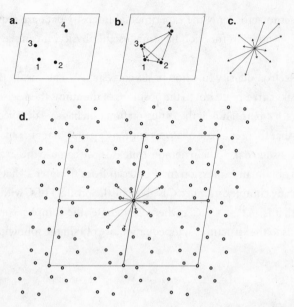

Figure 5.1. Building up the Patterson function from the atomic positions. (a) One unit cell of a simple two-dimensional crystal containing four atoms. (b) The same unit cell shown in (a), but showing the vectors connecting the four atoms. (c) The interatomic vectors from (b), shown emanating from a common origin. This is how the vector peaks appear in the Patterson function. (d) Multiple unit cells of the Patterson function. Peaks corresponding to interatomic vectors are shown as dots. The vectors are drawn for only the central unit cell, but all the unit cells are identical.

If we use F as the coefficient in a Fourier series we obtain the electron density, $\rho(x)$. $\rho(x)$ has peaks corresponding to the atomic positions, x_j. By analogy, we might expect that using F^2 as the coefficient in a Fourier series will give us a function, $P(u)$, that has peaks corresponding to the interatomic vectors, $(x_j - x_k)$. This is indeed the case and is easily demonstrated. Combining equations (5.1) and (5.2) we obtain

$$P(u) = \frac{1}{V} \sum_h \left[\sum_j \sum_k f_j f_k \exp(2\pi i h \cdot [x_j - x_k]) \right] \exp(-2\pi i h \cdot u)$$

Combining terms leads to the following:

$$P(u) = \frac{1}{V} \sum_h \sum_j \sum_k f_j f_k \exp(-2\pi i h \cdot [u - (x_j - x_k)])$$

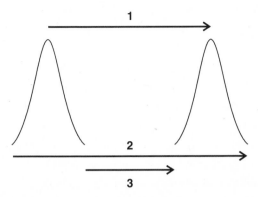

Figure 5.2. The Patterson peak corresponding to the vector between two atoms is roughly twice as broad as either of the atoms themselves. The figure shows the profiles through the centers of two atoms. The Patterson peak will contain not only vector 1, running between the centers, but all possible vectors running from any point in one atom to any point in the other. These include the long vector 2, running from the trailing edge of the first atom to the leading edge of the second, and the short vector 3, running between the leading edge of the first atom and the trailing edge of the second. The difference in length between vectors 2 and 3 defines the width of the Patterson peak and is approximately twice the width of the atoms.

At what values of u does this expression for $P(u)$ take on large values? In other words, where are the peaks in $P(u)$? It is clear that $P(u)$ will be maximized when the argument of the exponential equals zero; this is true when $u = x_j - x_k$. Hence, peaks occur at the positions u that correspond to interatomic vectors.

For every pair of atoms in the original molecule, the Patterson function contains a peak corresponding to the vector connecting these two atoms. In fact, there will be two vectors: $u_{1 \to 2}$ corresponds to the vector from atom 1 to atom 2, and $u_{2 \to 1}$ corresponds to the vector from atom 2 to atom 1. $u_{1 \to 2} = -u_{2 \to 1}$. This means that the Patterson function is centrosymmetric: $P(u) = P(-u)$. The relation between atomic positions, interatomic vectors, and the Patterson function is illustrated in Figure 5.1.

Comparison of equations (5.2) and (5.3) suggests that other differences exist between the electron density function and the Patterson function. For example, the shapes of the peaks in the electron density function are determined by the atomic scattering factors f_j, and the shapes of the peaks in the Patterson function are determined by the product of atomic scattering factors $f_j f_k$. Note that this product is effectively twice the width of the individual atomic scattering factors, so Patterson peaks are twice as broad as atoms (Figure 5.2).

Another difference between $\rho(x)$ and $P(u)$ is the number of peaks. If there are N atoms in the unit cell, there are N peaks in $\rho(x)$, but there are N^2 peaks in the Patterson function. The N Patterson peaks for which $j = k$ (*self-peaks*) are superimposed at the origin. Thus $P(u)$ always has a huge peak at the origin.

Because there are so many peaks in the Patterson function, $P(u)$ represents many superimposed interatomic vectors u. $P(u)$ can be thought of as the relative probability that two scattering centers in the specimen are separated by the vector u. Such a probability density is called a *pair correlation function*. Pair correlation functions arise in many circumstances when the Fourier transform of a distribution of scattering intensity is taken. For example, in solution scattering, where all possible molecular orientations are present, one can obtain a pair correlation function that gives the probability that two scattering centers are separated by a distance u, without regard to direction.

5.2 Using the Patterson Function to Locate Atoms

We can often work backward from the set of Patterson vectors to determine the actual positions of the atoms. There are many examples in the history of small-molecule crystallography. One illustrative special case occurs when a small organic molecule contains a single heavy atom. In that case the vectors between the heavy atom and the light atoms are actually position vectors, and the Patterson function provides a (noisy) image of the structure itself with the heavy atom placed at the origin. This is illustrated in Figure 5.3.

Macromolecular Patterson functions have too many peaks for such methods to work. For example, if a protein crystal has 10,000 atoms in the unit cell (this is about average) then 100 million Patterson peaks are present in one unit cell of the Patterson function. Obviously, these peaks must extensively overlap; in fact, the Patterson function of a protein is blurred and featureless. How can anything useful be obtained from such a messy function?

The trick is to focus on heavy atoms. Vectors between heavy atoms give very strong peaks that stand out from the others. Equation (5.2) tells us that the height of the Patterson peak corresponding to the vector between atoms j and k is proportional to the product of scattering factors, $f_j f_k$. Recall that the magnitude of f is proportional to the number of electrons in the atom. Thus, a Patterson peak corresponding to a vector between two carbon atoms has a weight of $6 \times 6 = 36$, whereas a peak between two uranium atoms has a weight of $92 \times 92 = 8,464$, almost 300 times stronger than the carbon-carbon peak. For molecules contain-

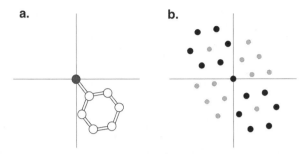

Figure 5.3. An example of how atomic positions can be inferred from the Patterson function when heavy atoms are present. (a) A simple organic compound—iodobenzene—containing a single heavy atom. The iodine atom is dark gray and is placed at the origin of an arbitrary coordinate system. (b) The Patterson function calculated from this molecule. Peaks in the Patterson function are shown as dots. Vectors between the iodine atom and carbon atoms appear as dark peaks, while carbon–carbon vectors are shown as light peaks. There is also a peak at the origin, as is always the case for the Patterson function. Note how the dark peaks reveal both the structure of the molecule and its mirror image.

ing a moderate number of light atoms plus a few heavy atoms, the heavy atom–heavy atom peaks can frequently be identified in the Patterson function, allowing the heavy atom positions to be inferred.

Unfortunately, in large molecules like proteins, the Patterson functions are so complex that even heavy atom peaks become lost. However, as we have seen in Chapter 4, the positions of the heavy atoms must be known before we can calculate phases in the MIR experiment. How do we find them? In this case, we use what are called difference Patterson functions to accentuate the heavy atom peaks and allow their identification.

Suppose we are searching for the positions of the heavy atoms in a heavy atom derivative of a protein. The heavy atoms are bound to the protein and occupy specific sites in the crystal lattice. Now imagine that we could magically erase all the protein atoms from the crystal, without changing the positions of the heavy atoms. The diffraction pattern of this imaginary heavy atom-only crystal would correspond to the structure factors $F_H(\boldsymbol{h})$. The Patterson function calculated from these data would have coefficients $F_H^2(\boldsymbol{h})$ and would be ideally suited for determining the positions of the heavy atoms. Of course, we can't erase the protein atoms, and we can't measure F_H, but we can approximate F_H. As mentioned in Chapter 4, the isomorphous difference $\Delta F_{\text{iso}} = F_{PH} - F_P$ can be used as an estimate of F_H. Using ΔF_{iso} we can calculate what is known as an *isomorphous difference Patterson function*:

$$P_{iso}(\boldsymbol{u}) = \frac{1}{V} \sum_{\boldsymbol{h}} (F_{PH}(\boldsymbol{h}) - F_P(\boldsymbol{h}))^2 \exp(-2\pi i \boldsymbol{h} \cdot \boldsymbol{u})$$

$$= \frac{1}{V} \sum_{\boldsymbol{h}} \Delta F_{iso}^2(\boldsymbol{h}) \exp(-2\pi i \boldsymbol{h} \cdot \boldsymbol{u}) \qquad (5.4)$$

This function is not identical with the true heavy atom Patterson calculated with $F_H^2(\boldsymbol{h})$ coefficients. However, it does contain strong peaks at positions corresponding to interatomic vectors for the heavy atoms. Figure 5.4 shows an example of an isomorphous difference Patterson function.

When using the difference Patterson function to locate heavy atoms in protein crystals, it is useful to exploit the symmetry of the crystal. For example, suppose the crystal contains a twofold rotation axis of symmetry along the y coordinate, as do crystals belonging to the monoclinic space group $P2$. If the unit cell contains a heavy atom at the point x, y, z, then it will contain a second heavy atom generated by the twofold symmetry axis at position $-x, y, -z$. These are called equivalent positions. The difference Patterson vector between these two heavy atoms is $x - (-x), y - y, z - (-z) = 2x, 0, 2z$. Also, because the Patterson function is centrosymmetric, a second peak will occur at $-x - x, y - y, -z - z = -2x, 0, -2z$. If we examine only the $y = 0$ section in the Patterson function, we will find the peaks corresponding to the vectors between the two equivalent, symmetry-related heavy atoms. We know that the peaks are located at $\pm 2x, \pm 2z$, so by measuring the positions of the peaks we can obtain the x and z coordinates of the heavy atom. Peaks representing vectors between symmetry-related heavy atoms are major features in difference Patterson functions. The special sections that contain vectors between symmetry-related atoms (like the $y = 0$ section in the preceding example) are called *Harker sections*, after the protein crystallography pioneer David Harker. See the legend of Figure 5.4 for a little more information on determining the heavy atom positions from the Patterson function.

Because the Patterson function is centrosymmetric, it does not allow us to determine the absolute handedness of a set of atoms or a molecule. If one starts with the peak at $-\boldsymbol{u}$ then all of the coordinates derived will be the inverse of those derived using the peak at $\boldsymbol{u}$. We don't know in advance which peak to choose, so

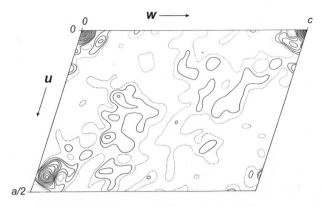

Figure 5.4. Isomorphous difference Patterson map for a mercury derivative of a protein. The crystals in this example contain hemoglobin from the annelid *Glycera dibanchiata*. This two-dimensional contour plot shows the $v = 0$ section of the full three-dimensional Patterson function (the u-w plane corresponds to the x-z plane in the real unit cell). The section plotted extends over the full unit cell in w (across) and halfway along the unit cell in u (down). The origin is at the upper left, and the large peak there represents vectors between atoms and themselves. The peak about halfway down on the left, lying at approximately $u = 0.45$ and $w = 0.04$, represents a vector between two symmetry-related mercury atoms. It occurs at the position $2x, 2z$, where x and z are the mercury atom coordinates in the unit cell. The refined coordinates for the mercury atom are $x = 0.225$ and $z = 0.007$. Reproduced from the dissertation *The Structure of Glycera Hemoglobin* by E. A. Padlan.

we may choose the correct arrangement of heavy atoms, or we may choose its inverse. If the incorrect arrangement of heavy atoms is chosen, it will give rise to an inverse image—a protein containing D-amino acids and left-handed α-helices. Changing the sign of every phase angle will invert the handedness of the image, and most programs contain a switch to do this. (Crystallography *is* capable of determining the absolute handedness of molecules through the use of anomalous scattering, but we won't describe how in this book. Note that Linus Pauling, Robert Corey, and Herman Branson, in their landmark 1951 paper describing the structure of the α-helix, did not attempt to assign absolute handedness and actually chose arbitrarily to draw the helix as left-handed! Later that same year Johannes Bijvoet published his account of how the absolute configuration of chiral molecules can be determined by using anomalous scattering.)

The example we just examined offers no information about the y coordinate of the heavy atom. In this case, we are at liberty to set the y coordinate equal to zero. This point is not at all obvious, but it follows from the rules that crystallographers have established for choosing the origin of the unit cell. Generally speaking, the origin is chosen at a point that bears some simple relation to the symmetry elements of the unit cell. Thus, for space group $P2$, the convention is to choose the origin to coincide with the twofold rotation axis. Because the twofold axis runs parallel to the y axis, this severely limits the origin choice in the x-z plane. However, any point along y may be chosen as the origin.

Obviously, if there are two distinct heavy atom binding sites in this derivative, we can set the y coordinate of only one of them equal to zero. The y coordinate of the other site is determined from Patterson vectors that run from the first heavy atom site to the second. Suppose that the heavy atom bound to site 1 has coordinates x_1, 0, z_1, and the atom at site 2 has coordinates x_2, y_2, z_2. The Patterson vector between these atoms is given by $u = x_2 - x_1$, $v = y_2 - 0 = y_2$, and $w = z_2 - z_1$. We can find the values of x_1, z_1 and x_2, z_2 by using Harker sections, as described above. This gives us u and w for the vector connecting the two sites. These values of u and w define a line in the Patterson function. By scanning along the line we will find a peak at some value of v, corresponding to y_2.

Crystals may contain rotation axes running in more than one direction. In such cases, the choice of origin is limited to a few values along x, y, and z (for example, the points where the axes intersect). In general, such space groups contain Harker sections perpendicular to two or more axes, allowing us to measure directly all three coordinates of a heavy atom.

If we have two distinct heavy atom derivatives, say mercury and lead, we do not initially know if our choices of origin for the two derivatives are the same. Trial and error is used to find a common choice. In MIR the phase circle construction produces a point of intersection of the three circles. If the two heavy atom derivatives do not share a common origin, the circles will generally not intersect as they are supposed to. For this reason, programs that calculate phases test different combinations of origin choices and choose the one that gives the most consistent pattern of circle intersections. We do not need to worry about whether the choices of origin we make for the native protein and heavy atom derivatives are the same, since the phases of the native crystal are determined by the phases of the heavy atom derivatives. The native crystal is constrained to have a common origin with the heavy atom derivatives.

Isomorphous difference Patterson functions allow us to determine the positions of the heavy atoms used in the MIR method. What about the anomalous scatterers used in the MAD method? In this case we can use the anomalous difference $\Delta F_{\text{anom}} = F(h) - F(-h)$. The anomalous difference Patterson function

$$P_{\text{anom}}(u) = \frac{1}{V} \sum_h (F(h) - F(-h))^2 \exp(-2\pi i h \cdot u) \qquad (5.5)$$

contains peaks that correspond to the vectors connecting anomalous scatterers in the structure. We do not need to be concerned about consistent origin choice for the MAD method, because all of the data are measured from a single crystal and the positions of the anomalous scatterers do not change with wavelength.

5.3 Summary

- The Patterson function $P(u) = \frac{1}{V} \sum_h F^2(h) \exp(-2\pi i h \cdot u)$ can be calculated directly from the experimental data and requires no knowledge of the phases.
- The Patterson function is defined in the same unit cell as the electron density but contains peaks at points $u_{jk} = (x_j - x_k)$, where x_j and x_k are the positions of any two atoms. Thus, while peaks in the electron density function correspond to atomic positions, peaks in the Patterson function correspond to vectors connecting atomic positions. A structure with N atoms will have N peaks in the electron density function and N^2 peaks in the Patterson function.
- The Patterson peak corresponding to the vector between atoms j and k has a weight proportional to the product of the number of electrons in atoms j and k. Therefore, Patterson vectors between heavy atoms are very strong.
- Heavy atom positions can often be found by working backward from the (relatively small) set of heavy atom Patterson peaks. However, in large molecules like proteins the Patterson functions are so crowded that it is impossible to find even these heavy atom peaks. For such molecules we use difference Patterson functions to make the heavy atoms stand out more clearly. In the difference Patterson function, F^2 is replaced by ΔF^2, where ΔF can be either the isomorphous difference $\Delta F_{\text{iso}} = F_{PH}(h) - F_P(h)$ or the anomalous difference $\Delta F_{\text{anom}} = F(h) - F(-h)$.

FURTHER READING

The Patterson function is covered particularly well in Chapter 12 of *X-ray Structure Determination* by G. H. Stout and L. H. Jensen (John Wiley and Sons, New York, 1989). This book is in print but used copies present a good value.

Crystal Structure Analysis: A Primer. 2nd edition, by J. P. Glusker and K. N. Trueblood (Oxford University Press, New York, 1985), referenced earlier, provides a fine concise treatment.

The International Union for Crystallography sponsored a symposium in 1984 entitled *Patterson and Pattersons. Fifty years of the Patterson function* (IUCr Crystallographic Symposia, Vol. 1.; eds. Jenny P. Glusker, Betty K. Patterson, and Miriam Rossi; New York, Oxford University Press, 1987). This volume contains examples of the theory and practice of the Patterson function, as well as personal reminiscences about A. L. Patterson, which help reveal the human face of crystallography. Out of print, but look for it in your university library.

6

Phasing with Partially Known Structures

6.1 Difference Fourier Maps

Often we want to know the crystal structure of a molecule when the structure of a similar one is already known. In fact, this is by far the most common problem that crystallographers face. For example, one may wish to determine the structure of a mutant of a particular protein, when the structure of the wild type is already known.* Other common cases include proteins that are crystallized in the presence and absence of ligand or proteins that are evolutionary relatives of known proteins. This chapter discusses methods that make use of known structural information to simplify the problem of structure determination. For convenience, we refer to the molecule whose structure we'd like to find as the *target* molecule and to the known structure as the *reference* molecule.

In particular, when the crystals of the target and reference molecules are isomorphous the problem is straightforward. This circumstance is most common when the target is very similar to the reference—proteins differing at a single amino acid position, for example, or proteins crystallized in the presence or ab-

---

*For the most part, small changes in a protein's sequence create only local structural changes, so it's usually safe to assume that the overall structure of the mutant molecule will be similar to that of the wild type. For example, hundreds of structures are known for variants of the antigen-binding antibody fragment called Fab; all of them share the same overall fold and differ only in the residues at the antigen binding site.

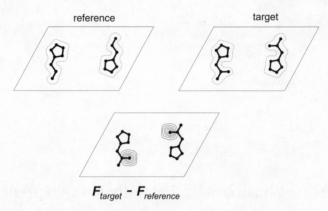

Figure 6.1. A cartoon depicting a simple case in which knowledge of the structure of one molecule can be used to determine the structure of a closely related molecule (provided the crystals of the two molecules are isomorphous with one another). In the upper left is shown the known crystal structure of the reference molecule; at upper right is shown the unknown structure of the isomorphous target crystal. The target molecule differs from the reference in having one additional carbon atom. A difference Fourier map calculated using the coefficients $F_{target} - F_{reference}$ and phases calculated from the reference structure is shown below. It contains only two peaks, corresponding to the extra carbons on the two copies of the target molecule.

sence of a ligand. Figure 6.1 illustrates such a case for a simple two-dimensional crystal.

This situation is similar to the one we face in refinement, where the initial model built into the electron density has inaccuracies that must be rectified. (Refinement is discussed in more detail in Chapter 7.) We can thus regard the reference structure (the structure for which we know the atomic coordinates) as a preliminary or unrefined version of the target structure (the structure for which we have measured the X-ray data). Any appropriate refinement method may be used, but we will see that difference Fourier methods are often a good way to start.

Difference electron density maps reveal structural differences between two isomorphous crystals. Imagine we have two isomorphous crystals of a protein— say, one with a small molecule ligand bound and one without. The ideal way to visualize the differences between these two structures would be to subtract the electron density function of one from that of the other. The resulting function would show only the differences between the two structures. However, to calcu-

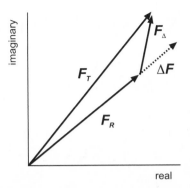

Figure 6.2. Vector diagram showing the relation between the diffraction of the reference and the target in a difference Fourier map. F_Δ reflects the true difference between the two structure factors, but we cannot measure F_Δ directly. We therefore approximate F_Δ by the dotted vector ΔF and use $|\Delta F|$ as the difference map coefficient. This appears to be a poor approximation, because the angle between ΔF and F_Δ (the "phase error") can be quite large. Fortunately, as the phase error increases, the amplitude $|\Delta F|$ decreases, so that when the phase error is at its worst—90°—the amplitude falls to zero and this reflection makes no contribution to the map.

late such an ideal image we would need to know the phases for both structures. The difference map (often called a difference Fourier) is an approximation to the ideal image that does not require phases for the target structure. Difference maps are typically flat and featureless throughout most of the unit cell, where the molecules are identical, but they show peaks (positive or negative) where one molecule has a feature that the other lacks. The difference map is calculated by the relation

$$\Delta\rho(\boldsymbol{x}) = \frac{1}{V} \sum (F_T(\boldsymbol{h}) - F_R(\boldsymbol{h}))\, e^{i\alpha_R} e^{-2\pi i \boldsymbol{h}\cdot\boldsymbol{x}} \tag{6.1}$$

where F_T and F_R are the structure factor amplitudes from the target and reference crystals, respectively. An approximation made in this equation is that the phases from the target and reference crystals are the same. Since we do not know the phases from the target, we use the reference phase (α_R) for both. Because the structures are similar, one might assume that this is an excellent approximation, but a more careful look reveals some problems.

Figure 6.2 shows the true difference coefficient, labeled $F_\Delta = F_T - F_R$, and

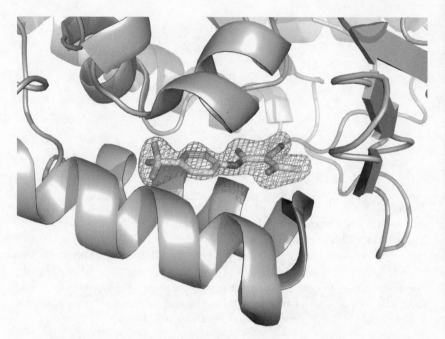

Figure 6.3. An $F_{obs} - F_{calc}$ difference Fourier map showing the binding of an inhibitor to an enzyme. The enzyme shown is derived from the malaria parasite *Plasmodium falciparum*, and the inhibitor represents a so-called lead compound for the development of an antimalarial drug. The inhibitor structure has been fitted into the density of the difference map. The values for F_{calc} and the phases were calculated from a model of the protein with no inhibitor, while the F_{obs} values represent the diffraction data measured from the crystal of the protein bound to the inhibitor. This structure can be found in the Protein Data Bank under accession number 1tv5.

how it is related to ΔF, which is the approximate coefficient used in equation (6.1). Clearly, ΔF can differ markedly from F_Δ. In general, $|\Delta F| \neq |F_\Delta|$, and the phase of F_Δ is random compared with the phase of ΔF. ΔF therefore appears to be a poor approximation for F_Δ. We have alluded to this problem before—it appears when we use the isomorphous differences to approximate heavy atom scattering (Chapters 4 and 5).

Fortunately, despite these problems, it is possible to show that structural differences are represented accurately in the difference Fourier map (we won't include the proof here). The price we pay for the ΔF approximation is this: although the correct features do appear in the difference maps, the peaks corresponding to these features are only half the height they would be in an error-free map. In addition, noise is introduced into the maps by the approximation. As long as the differences between the structures in the two isomorphous crystals are small, however, the noise peaks are small compared with the true features.

In practice, here is how the difference Fourier would be used for a structure determination of a protein:ligand complex. First, the structure of the protein alone is determined. Next, diffraction data are collected from isomorphous crystals of the protein containing the bound ligand. A difference Fourier map is then calculated using the coefficients $(F_{obs} - F_{calc})$ and phases α_{calc}, where F_{obs} is the observed structure factor amplitude from the complex crystals, and F_{calc} and α_{calc} are the structure factor amplitude and phase calculated from the structure of the uncomplexed protein. An example of such an "$F_o - F_c$" map is shown in Figure 6.3. This map demonstrates how the location of the drug in a drug:protein complex can be revealed by a difference Fourier synthesis in which the only phase information is derived from the uncomplexed protein.

6.2 Molecular Replacement

When the crystals of the target and reference molecule are not isomorphous the problem is messier. We no longer have a preliminary model that is almost correct, as we would if the two crystal forms were isomorphous. We still know that the structure of the target molecule will be similar to that of the reference, but we do not know where the molecule is in the unit cell. The problem therefore boils down to correctly orienting and positioning the reference molecule in the target's unit cell. (A particular choice for the position and orientation of a molecule is sometimes called a *pose*.) Once this is accomplished, the resulting model can be improved using the refinement methods discussed in Chapter 7. The methods used to position the reference molecule have been termed *molecular replacement*.

The reference molecule is positioned by a search procedure—essentially a sophisticated trial-and-error process. Figure 6.4 shows a target structure and several possible models for it, created by placing the reference molecule in the target unit cell in different orientations and positions. We can easily see that most of these guesses are wrong. However, when actually determining a structure we do not know the correct answer in advance; how then would we know that these models were wrong? The test is to calculate structure factors from all these models and to compare them with the observed structure factor amplitudes measured from the target crystal. For the incorrect guesses, the agreement is poor, signaling that the models are not satisfactory representations for the target crystal structure; agreement should be better for the correct guess.

The effect of rotating a molecule on its diffraction pattern is illustrated in

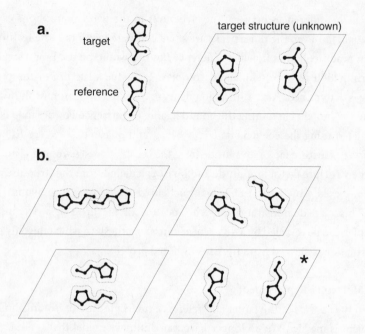

Figure 6.4. The process of molecular replacement. (a) We use the same target and reference molecules as in Figure 6.1, but we no longer assume that the crystals of the target and reference are isomorphous. Thus, although we know the structure of the reference molecule (and assume that the structure of the target molecule is similar), we do not know how the molecule is positioned within the unit cell of the target crystal. (b) Cartoons showing possible models for the target structure, obtained by placing the reference molecule in different orientations and positions within the target unit cell. Three of the four guesses shown are clearly wrong (the most correct guess is marked with an asterisk). Note that some of the guesses actually cause symmetry-related molecules in the cell to overlap, which is physically impossible. This is a common occurrence when we attempt brute-force solutions, in which all possible orientations and positions are tested. The presence of such bad contacts can help weed out incorrect answers when we are assessing several potential solutions.

Figure 6.5. The structure of the molecule does not change when it is rotated and therefore neither does its diffraction pattern. However, the diffraction pattern does rotate along with the molecule. Recall that the diffraction pattern of a crystal represents a sampled version of the molecule's diffraction pattern (this point is illustrated by Figure 3.6). Because of this sampling effect, as the reference molecule is rotated the intensities of different reflections change significantly. We wish to orient the reference molecule so that the predicted and observed intensities match.

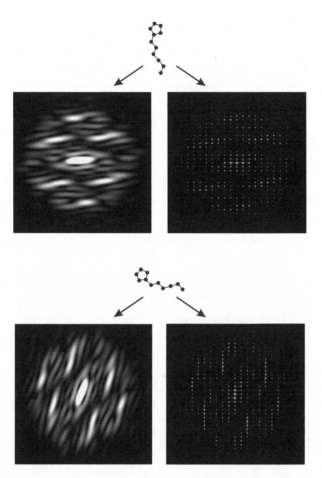

Figure 6.5. The effect of rotating an object upon its diffraction pattern. Two orientations of the same molecule are shown. For each orientation, the Fourier transform of the single molecule is shown on the left, and the Fourier transform of a crystalline array of that molecule is shown on the right. Rotating the molecule rotates its diffraction pattern, and the crystal diffraction pattern represents a sampled version of the molecular diffraction pattern. We know the unit cell parameters for the target crystal —these define the spacing of the spots in the diffraction pattern, or, if you will, the mask through which we observe the molecule's diffraction pattern. In molecular replacement, we are holding the crystal mask fixed and rotating the molecular transform underneath. The correct orientation is the one in which the pattern of spot intensities as seen through the mask best matches the observed diffraction intensities.

In molecular replacement, a systematic search procedure is used to sample all possible orientations and positions of the reference molecule in the target unit cell. The orientation and position for which the calculated and observed structure factors agree most closely provide a plausible guess for the answer. Elaborate software packages exist for carrying out these searches, but a detailed analysis of their workings is beyond the scope of this book. However, the basic ideas are quite straightforward.

An exhaustive, six-dimensional search is the obvious (if naïve) choice:

- Place the reference molecule successively at each of the points on a fine grid sampling the asymmetric unit of the target crystal unit cell.
- At each test position, place the reference molecule into all possible orientations, again using a fine grid, creating a set of poses.
- For each pose, generate the symmetry mates of the reference molecule. This produces a possible model for the crystal of the target molecule.
- Calculate $F_R(\boldsymbol{h})$ for each potential model, and compare it with $F_T(\boldsymbol{h})$.

This comparison can be made by calculating the correlation coefficient between $F_R(\boldsymbol{h})$ and $F_T(\boldsymbol{h})$, or by calculating the R value:

$$R = \frac{\sum_{h} |F_R(\boldsymbol{h}) - F_T(\boldsymbol{h})|}{\sum_{h} F_T(\boldsymbol{h})} \qquad (6.2)$$

The model with the highest correlation coefficient and/or lowest R value corresponds to the correct solution (at least in principle).

This is a six-dimensional search. We have to scan three angles and three coordinates in rather fine steps. Even with today's powerful computational capabilities, this is a daunting challenge. To simplify the calculation, Michael Rossmann and David Blow had the idea of separating the rotational and translational searches, performing two sequential three-dimensional searches instead of one six-dimensional search. The idea behind their method can be seen most easily using the Patterson function.

6.2.1 We Discuss Rotational Searches First

Peaks in the Patterson function correspond to vectors between pairs of atoms in the crystal. Some vectors will connect atoms within a single molecule and are called self-vectors. Others will connect atoms in two different molecules—cross vectors. Most of the time, pairs of atoms within a single molecule will be closer to each other than two atoms in two different molecules. *Thus, self-vectors in the Patterson function tend to lie near the origin, and cross vectors away from the origin.* When a protein molecule is rotated, the distances between atoms within that molecule do not change. Therefore, rotation of a molecule changes the directions of the self-vectors in the Patterson function but not their lengths. Herein lies our recipe for molecular replacement: For all possible orientations of the reference molecule, calculate the Patterson function. Next, compare the self-vectors from all these Patterson functions with the self-vectors from the Patterson function of the target crystal. The best agreement will occur when the reference molecule is in the same orientation as the molecule in the target crystal. We do this in practice by using the correlation function $R(\theta_1, \theta_2, \theta_3)$, which is known as the *rotation function.*

$$R(\theta_1,\theta_2,\theta_3) = \int_{\substack{\text{all} \\ \text{space}}} P_{\text{reference}} \left(M(\theta_1,\theta_2,\theta_3) \cdot u\right) P_{\text{target}} \left(u\right) dV \qquad (6.3)$$

Here M is a matrix that carries out the rotations and is a function of three rotation angles θ_1, θ_2, θ_3. The function $P_{\text{reference}}$ is the Patterson function for the reference molecule that has been edited to contain only self-vectors. $P_{\text{reference}}$ falls to zero past a radius whose length is equal to that of the longest vector in the molecule. Thus, although the integral is written over all vector space, the integrand is only nonzero in a limited volume.

The rotation function works because the integral is large when peaks in $P_{\text{reference}}$ fall on top of peaks in P_{target}. Therefore, the trio of angles θ_1, θ_2, θ_3 for which R has the largest value should describe the orientation of the reference molecule in the unit cell of the target crystal.

The goal of the rotation function is to orient molecular models within the unit cell in ways that agree with the observed diffraction data. This involves rotating our model. How do we do this? Rotations may be described using angles devised by Euler in the eighteenth century. According to Euler's rotation theorem, any rotation may be described by using successive rotations about three

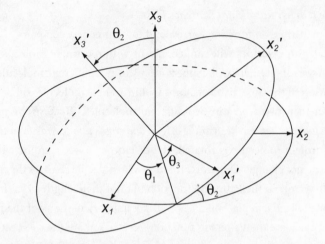

Figure 6.6. The three successive rotations that comprise a general rotation using Euler angles.

coordinate axes embedded in the object being rotated. If the three axial rotations are written in terms of rotation matrices B, C, and D, then a general rotation A can be written as

$$A = BCD \tag{6.4}$$

The three angles giving rise to the three rotation matrices are called Euler angles. Several conventions for Euler angles exist, depending on how we draw the axes about which the rotations are performed. The so-called x convention, illustrated in Figure 6.6, is the most common definition. In this convention, the rotation specified by Euler angles $(\theta_1, \theta_2, \theta_3)$ is accomplished by first rotating by an angle θ_1 about the z axis, then by an angle θ_2 about the x axis, and third, by an angle θ_3 about the z axis (again). Note, however, that several other conventions in which the rotations take place about other axes are also in common use.

In the x convention, the component rotations are given by

$$B = \begin{bmatrix} \cos\theta_3 & -\sin\theta_3 & 0 \\ \sin\theta_3 & \cos\theta_3 & 0 \\ 0 & 0 & 1 \end{bmatrix}$$

$$C = \begin{bmatrix} 1 & 0 & 0 \\ 0 & \cos\theta_2 & \sin\theta_2 \\ 0 & \sin\theta_2 & \cos\theta_2 \end{bmatrix}$$

$$D = \begin{bmatrix} \cos\theta_1 & \sin\theta_1 & 0 \\ -\sin\theta_1 & \cos\theta_1 & 0 \\ 0 & 0 & 1 \end{bmatrix}$$

Multiplying these three matrices together to obtain A we find

$$A = \begin{bmatrix} \cos\theta_3\cos\theta_1 - \cos\theta_2\sin\theta_1\sin\theta_3 & \cos\theta_3\sin\theta_1 + \cos\theta_2\cos\theta_1\sin\theta_3 & \sin\theta_3\sin\theta_2 \\ -\sin\theta_3\cos\theta_1 - \cos\theta_2\sin\theta_1\cos\theta_3 & -\sin\theta_3\sin\theta_1 + \cos\theta_2\cos\theta_1\cos\theta_3 & \cos\theta_3\sin\theta_2 \\ \sin\theta_1\sin\theta_2 & -\sin\theta_2\cos\theta_1 & \cos\theta_2 \end{bmatrix}$$

A useful way to think about the matrix operator A is to imagine it operating on the coordinate axis system x, y, z to produce a new, rotated version of the coordinate system. If x is a vector in the fixed (unrotated) coordinate system, it will have another name, say x', in the rotated system. These two names for the same vector are related by

$$x' = Ax$$

The rotation matrix M used in equation (6.3) will work exactly like the matrix A described above. Imagine that the functions P_{target} and $P_{reference}$ from equation (6.3) are superimposed in space and that you are focusing on a particular point x in P_{target}. We then rotate $P_{reference}$ with respect to P_{target}. As we rotate $P_{reference}$, different points in that function will fall on top of our chosen point x. For each orientation of the reference molecule, Mx is the point in $P_{reference}$ that falls on top of the point x in P_{target}. Thus, M is a tool for relating the rotated and unrotated Patterson functions.

The other major system for representing rotations makes use of *spherical polar angles*, which are more intuitive than Euler angles. They are based on the theorem that any rotation can be accomplished by a single spin about a properly chosen axis. Thus, two angles are used to specify the longitude and colatitude of the axis, and the third angle is used to specify the spin around it. This angular

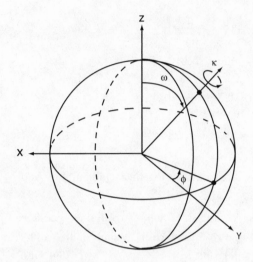

Figure 6.7. Spherical polar angles. Two angles specify the longitude and colatitude of the rotation axis, and one angle specifies the spin about that axis. These angles are particularly useful for the self-rotation function, where one may expect to find simple rotations such as twofold axes. A section of the rotation function at $\kappa = 180°$ will show peaks on the sphere corresponding to twofold axes.

system is shown in Figure 6.7. The matrix that specifies rotations in the spherical polar angle system is rather messy. It can be found in the original Rossmann and Blow paper.[*]

The use of Euler angles has its difficulties. If the value of the second Euler angle is small, the first and third rotations are highly correlated, because they are made around nearly parallel axes. This results in highly distorted peaks in the rotation function. Use of modified (or quasiorthogonal) Euler angles can alleviate this problem. The modified Euler angles are referred to as θ_+, θ_-, and θ_2 and are given by:

$$\theta_+ = (\theta_1 + \theta_3)/2$$
$$\theta_- = (\theta_1 - \theta_3)/2$$
$$\theta_2 = \theta_2$$

In Figure 6.8 a contour plot is shown for a two-dimensional rotation function. Figure 6.8a uses conventional Euler angles, and Figure 6.8b uses the modi-

―――――
[*]Rossmann, M. G., and Blow, D. M. *Acta Crystallogr.* 1962;15:24.

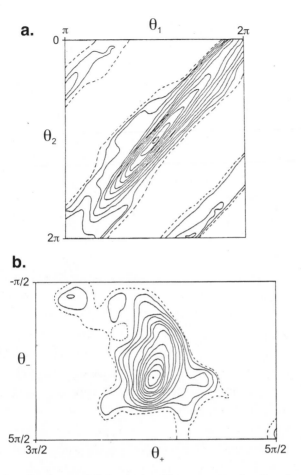

Figure 6.8. Quasi-orthogonal Euler angles. The same rotation function section is calculated by using Euler angles (a) and quasi-orthogonal Euler angles (b).

fied ones. Note that the peak appears much more symmetric when represented using the modified angles. When peak shapes are elongated and distorted, it is difficult to accurately measure the areas under these peaks. The more symmetric peaks obtained with the modified angles allow the weights of peaks in different regions of the rotation function to be quantitatively compared. These angles are incorporated as an option in many programs.

In this discussion, we've assumed that the rotation function will be used to orient a reference molecule in the first step of a molecular replacement structure determination. When Rossmann and Blow first conceived the rotation function in the early 1960s, however, a grand total of only two protein structures were known, and the notion of using known protein structures to solve unknown crystal structures was probably not foremost in their minds. In fact, the rotation function was originally developed with another purpose, that of revealing *noncrystallographic symmetry* within the crystal. What is noncrystallographic symmetry? Recall that for molecules to form a crystal, they must pack together in a way that obeys the symmetry of that particular crystal's space group. However, molecules sometimes possess additional symmetry that is not reflected by the crystal symmetry. This type of symmetry is known as noncrystallographic or local symmetry. For example, suppose we wish to crystallize a molecule that forms a symmetric C2 dimer—two identical monomers related by a twofold rotational axis of symmetry. The dimer might happen to crystallize in a space group containing a twofold symmetry axis, with the dimer's twofold coincident with the crystallographic twofold axis; in this case the crystal's asymmetric unit would contain one monomer. Alternatively, the molecule might crystallize in a different space group that contains a complete dimer in the asymmetric unit. In the latter case, although the dimer's twofold symmetry axis still exists, it is not coincident with any crystallographic twofold axis, and in fact the dimer symmetry axis is irrelevant to the symmetry of crystal packing. As another illustration, we note that many protein oligomers (and some viruses) contain fivefold axes of symmetry. Crystal lattices cannot contain fivefold axes, so the symmetry of the oligomers cannot be reflected in the symmetry of the crystal. The fivefold axis is therefore another example of noncrystallographic symmetry—it exists within the oligomer, but it is not used to build up the crystal lattice.

The rotation function is very useful for finding and characterizing noncrystallographic symmetry within a crystal. For this purpose we use a special form of the function known as the self-rotation function. In the self-rotation function, the Patterson function of the crystal is rotated and multiplied by an unrotated version of itself. When the angle by which we rotate the Patterson corresponds to the angle relating two subunits in the oligomer, then the rotated and unrotated Patterson are similar, and the integral of their product is large. The self-rotation function is plotted as a function of the rotation angles corresponding to all possible relative orientations. Peaks in the self rotation function correspond to the angles relating subunits, and reveal the direction of the noncrystallographic symmetry axes in the crystal.

An example of the self-rotation function is shown in Figure 6.9. The crystal

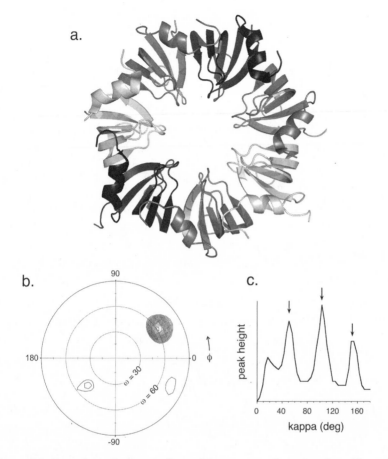

Figure 6.9. Detection of noncrystallographic symmetry by using the self-rotation function. (a) Structure of a protein isolated from the thermophilic microbe *Methanobacterium thermautotrophicum.* This protein forms a homoheptamer, with seven identical subunits arranged in a sevenfold symmetric ring. In this view, the sevenfold axis of symmetry is perpendicular to the plane of the page. The structure of this protein may be found in the Protein Data Bank under accession number 1jbm. (b) The $\kappa = 51°$ section of the self-rotation function. The self-rotation function is calculated as a function of the three polar angles ω, ϕ, and κ and is represented in a stereographic projection. In this projection the crystallographic *a* axis is perpendicular to the plane of the page; each section of the projection represents a single value of κ. Note the strong peak at $\phi = 33°$, $\omega = 62°$, revealing the direction of the noncrystallographic symmetry axis. (c) Graph of how the intensity of this peak varies with κ. Note how maxima occur at intervals of $51.4°$ ($360°/7$), reflecting the presence of sevenfold rotational symmetry in the crystal.

used for this figure contains a heptameric protein, in which the seven identical subunits are related by sevenfold symmetry (rotations of 51.4° and multiples thereof). The rotation function is plotted by using spherical polar angles; Figure 6.9b shows a plot of the function for which the spin angle κ has been limited to 51.4°. The strong peak at $\phi = 33°$, $\omega = 62°$ reveals the orientation of the sevenfold axis. Of course self-rotation functions can be carried out equally well using the Eulerian angular system; however, maps produced from spherical polar angles provide very direct visual pictures of the molecular symmetry.

6.2.2 Translational Search

Once the orientation of the reference molecule has been predicted by the rotation function, a translational search lets us find a complete model for the target crystal. We explore the translation function using an example with only two molecules in the unit cell to avoid the messy formalism required for the general case. Suppose that we have already determined the orientation of both molecules using the rotation function. Using this information, the translation function creates models of the target crystal structure and compares the diffraction patterns of these models with the observed data from the target crystals.

The structure factor for the entire crystal is the sum of the structure factors for each of the two component molecules.

$$F(h) = F_1(h) + F_2(h)$$

where the subscripts denote the two molecules. The diffraction intensity from the crystal is given by $|F(h)|^2$. What is the neatest way to calculate this? Recall from Chapter 5 that $|F(h)|^2 = F(h) \cdot F^*(h)$, where F^* is the complex conjugate of F.

To calculate the diffraction intensity expected from the crystal, we can write

$$|F|^2 = F(h) \cdot F^*(h) = (F_1 + F_2) \cdot (F_1^* + F_2^*)$$
$$= |F_1|^2 + |F_2|^2 + F_1 \cdot F_2^* + F_1^* \cdot F_2$$

where we have omitted the argument h for clarity.
We can then write the Patterson function as

$$P(u) = \Sigma(|F_1|^2 + |F_2|^2 + F_1 \cdot F_2^* + F_1^* \cdot F_2)\exp(2\pi i h \cdot u)$$

This deconstruction breaks the Patterson function into four terms. The first represents vectors within molecule 1; the second, vectors within molecule 2; the third, vectors from molecule 2 to molecule 1; and the fourth, vectors from molecule 1 to molecule 2. Because we are interested in determining the locations of our two molecules, the parts of the Patterson function that interest us are the third and fourth terms, which contain positional information. To exploit this positional information, we abstract the third term and define a "cross Patterson" function containing only intermolecular vectors:

$$P_{cross}(\boldsymbol{u}) = \frac{1}{V} \Sigma F_1(\boldsymbol{h}) \cdot F_2^*(\boldsymbol{h})\cos(2\pi\boldsymbol{h} \cdot \boldsymbol{u})$$

Note that the fourth term would have served equally well. The correlation function we wish to evaluate is analogous to the rotation function, and is given by

$$T = \int P_{cross}(\boldsymbol{u}) \, P_{xtal}(\boldsymbol{u})dV \tag{6.5}$$

We can also write this as

$$T = \Sigma F_1(\boldsymbol{h}) \cdot F_2^*(\boldsymbol{h}) \cdot |F_{obs}(\boldsymbol{h})|^2 \tag{6.6}$$

But how do the positions of molecules 1 and 2 enter into the calculation? Remember that F_1 and F_2 are calculated for the two reference molecules, centered at the origin. We are now testing different positions for these molecules in the unit cell to find their correct locations. When we move molecules 1 and 2 to positions $\boldsymbol{x}_1$ and $\boldsymbol{x}_2$, the diffraction patterns are phase shifted and become

$$F_1 \exp(2\pi i\boldsymbol{hx}_1) \text{ and } F_2 \exp(2\pi i\boldsymbol{hx}_2),$$

respectively. Plugging this into the equation for T gives

$$T(\boldsymbol{x}_1, \boldsymbol{x}_2) = \Sigma F_1(\boldsymbol{h})\exp(2\pi i\boldsymbol{hx}_1) \cdot F_2^*(\boldsymbol{h})\exp(-2\pi i\boldsymbol{hx}_2) \cdot |F_{obs}(\boldsymbol{h})|^2$$

$$T(\boldsymbol{x}_1 - \boldsymbol{x}_2) = \Sigma |F_{obs}(\boldsymbol{h})|^2 \cdot F_1(\boldsymbol{h}) \cdot F_2^*(\boldsymbol{h})\exp(2\pi i\boldsymbol{h}[\boldsymbol{x}_1 - \boldsymbol{x}_2]) \tag{6.7}$$

This is a form that is often evaluated. It is simply a Fourier series like the electron density function. Note that the position of the molecules appears indirectly, in the form of the vector between molecules, as is seen in a Patterson function.

Suppose the two molecules are related by symmetry, specifically by a twofold rotation axis lying along the crystal's y axis. In this case the vectors x_1 and x_2 are symmetry related, with components x_1, y_1, z_1 and $-x_1$, y_1, $-z_1$. The vector difference $x_1\text{-}x_2$ is therefore $2x_1$, 0, $2z_1$. Thus in this particular case T becomes

$$T(x_1, z_1) = \Sigma \mid F_{\text{obs}}(h)\mid^2 \cdot F_1(h) \cdot F_2^*(h) \exp(2\pi i [2hx_1 + 2lz_1])$$

This discussion has given the flavor of how the translation function works, without going into too much detail. Many modifications have been made to the translation function to reduce noise. For example, the target Patterson function $F^2_{\text{obs}}(h)$ can be modified to remove short-range (intramolecular) vectors, which do not depend on position and only add noise to the calculation.

The classical translation function defined above is not the only method for positioning molecules that have been previously oriented using the rotation function. For example, one can simply move the correctly oriented molecule to all possible positions in the unit cell and, at each position, calculate the correlation coefficient or R value relating F_{obs} and F_{calc}.

In practice, for cases in which the reference and target molecules have significant differences, molecular replacement can take on an aspect of brute force. One uses not only the highest peak in the rotation function to describe the orientation, but may test the top 20 peaks. For each of these a full translation function analysis has to be carried out. At the end of the day one may have several different solutions with similar R values. The ultimate test of which solution is correct is refinement; the models that can be improved until their calculated structure factors agree with the observed data will be the correct ones. Sophisticated software packages can perform molecular replacement searches in a highly automated manner.

Molecular replacement fails totally from time to time. For example, vectors longer than the size of the molecule should not be included in the rotation function, because we wish to limit the calculation to self-vectors. However, in elongated and/or tightly packed molecules, it may be impossible to avoid including large numbers of cross-vectors, thereby increasing the noise level and "poisoning" the rotation function. Also, when multiple molecules are present in the crystal asymmetric unit, each individual molecule must be found separately. If many molecules are present, each molecule accounts for only a small fraction of the total scattering, and so its signal in the rotation function may be too low to identify among the noise peaks.

6.3 Summary

- Difference Fourier maps can reveal missing or misplaced bits of structure in a molecular model. These maps use coefficients $\Delta F = (F_{obs} - F_{calc})$ and phases α_{calc}, where F_{calc} and α_{calc} are calculated from the model. Peaks in such maps appear at half-weight.

- Molecular replacement uses the known structure of a reference molecule to determine the crystal structure of a target molecule of similar structure. The rotation function finds the orientation of the reference molecule in the target unit cell, and the subsequent translation function finds its position. This yields a model suggesting how the target molecule is posed in the unit cell, which can then be refined.

- The self-rotation function compares a crystal's Patterson function with itself and can reveal internal symmetry within the asymmetric unit (noncrystallographic symmetry).

FURTHER READING

The Molecular Replacement Method (ed. M. G. Rossmann; Gordon and Breach, New York, 1972) is an interesting collection of earlier papers on the rotation function and molecular replacement, including many that explain methods well.

Check out the short review by Michael Rossmann, *The Molecular Replacement Method* (*Acta Crystallogr.* A 1990;46(2):73–82). Very clear.

The program package AMoRe, which contains an elegant fast algorithm for the rotation function, is described in (Navaza, J. AMoRe: an automated package for molecular replacement. *Acta Crystallogr.* A 1994;50(2):157–163).

The program package Phaser implements sophisticated statistical methods to better evaluate the rotation and translation functions used in molecular replacement, as described in (Read, R. J. Pushing the boundaries of molecular replacement with maximum likelihood. *Acta Crystallogr.* D 2001;57(10):1373–1382).

The difference Fourier map is covered well in the text by Blundell and Johnson, *Protein Crystallography* (Academic Press, New York, 1976).

7

Crystallographic Refinement

7.1 Refinement Improves the Model

The product of a crystallographic experiment is a "model." The model is simply the x, y, and z coordinates for every atom in the molecule (plus a set of temperature factors or B values, if there are sufficient data). In an MIR or MAD experiment, this model is built by fitting the experimentally derived electron density map. In molecular replacement, the model is the structure of the reference molecule, oriented and positioned in the unit cell of the target crystal.

Initial models contain errors. These errors have many sources: experimental error (noise) in electron density maps, bias contributed by the individual who built the model, or, in molecular replacement, genuine differences between the target and reference molecules. The initial model can be improved, however, by a process known as *refinement*. Refinement is important not only because it improves the model but also because it eliminates the subjectivity inherent in human map fitting, so that structures of the same molecule determined in different laboratories will agree within experimental error. A simple example of the types of changes that occur during refinement is shown in Figure 7.1.

Here is the key insight into what makes refinement possible: we can calculate the diffraction pattern corresponding to any model and compare the calculated data with the observed data. Refinement is the process of systematically altering the model so that the observed and calculated data agree more and more closely. We have already met the exact mathematical relationship that connects the model

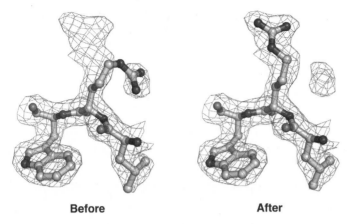

Before   **After**

Figure 7.1. Example of a portion of a crystallographic model before refinement (left) and after refinement (right). The models are superimposed on the electron density map. The side chain of the arginine residue is clearly misplaced in the initial model, which was derived from a molecular replacement structure. Figure courtesy Jason McLellan and Dan Leahy, Johns Hopkins School of Medicine.

with the diffraction data. It is the structure factor equation (3.11). Recall that the input to this equation is a set of atomic coordinates—a model—and the output is a set of $F(h, k, l)$.

7.2 Least-Squares Refinement

We now address how the model can be systematically improved. The simplest method is called *least-squares refinement*. This method is closely related to another least-squares process, namely that of fitting a straight line through a set of experimental data points. Comparing the two procedures is instructive.

Refinement and line fitting both have three requirements:

- A set of measurements of the independent and dependent variables. For a straight line these are the coordinates of the points $(x_{i,obs}, y_{i,obs})$; x is the independent variable, and y is the dependent variable. In crystallography, the independent variables are the Miller indices h, k, l, and the dependent variables are the $|F(h, k, l)|$.
- A mathematical model that relates the dependent variables to the independent variables. This allows you to calculate values for the dependent variable ($y_{i,calc}$ or $|F_{calc}(h, k, l)|$) and compare them with the observed values ($y_{i,obs}$ or

$|F_{obs}\,(h, k, l)|$). This model will contain several adjustable parameters. For a line the parameters are the slope and intercept, and for crystallography they are the atomic coordinates.

- A method of finding the values of these parameters that give the best fit of $y_{i,calc}$ to $y_{i,obs}$, or of $|F_{calc}\,(h, k, l)|$ to $|F_{obs}(h, k, l)|$.

For a line, the model we use is the following:

$$y_{i,calc} = mx_{i,obs} + b$$

where m and b are the parameters to be optimized.

Here is how least squares works. For a given m and b we can calculate a set of $y_{i,calc}$; in general, the $y_{i,calc}$ values do not equal the corresponding $y_{i,obs}$. We want the particular values of m and b that make $y_{i,calc}$ and $y_{i,obs}$ agree most closely. We can quantify the discrepancy between calculated and observed y values by defining Δy_i:

$$\Delta y_i = y_{i,obs} - y_{i,calc}$$

The model agrees best with the data when the absolute value of Δy_i, summed over all the data points, is at a minimum. Rather than minimize $|\Delta y_i|$, however, we choose to minimize the *squares* of Δy_i. Thus, we define $R = \sum_i (y_{i,obs} - y_{i,calc})^2$. Now the problem reduces to finding the values of m and b for which R is a minimum. To insert m and b into the equation we substitute for $y_{i,calc}$.

$$R = \sum_i (y_{i,obs} - y_{i,calc})^2 = \sum_i (y_{i,obs} - (mx_i + b))^2$$

We then minimize R by the usual expedient of setting its partial derivatives $\dfrac{\partial R}{\partial m}$ and $\dfrac{\partial R}{\partial b}$ equal to zero.

$$\frac{\partial (R)}{\partial m} = \frac{\partial \left(\sum_{i=1}^{n} (y_{i,obs} - y_{i,calc})^2 \right)}{\partial m} = 0$$

which gives

$$\sum_{i=1}^{n} -2(y_{i,\text{obs}} - mx_{i,\text{obs}} - b) \cdot x_{i,\text{obs}} = 0$$

On rearrangement this yields

$$m\sum x_{i,\text{obs}}^2 + b\sum x_{i,\text{obs}} = \sum x_{i,\text{obs}}y_{i,\text{obs}}$$

This expression is a linear equation in m and b. The terms under the summation signs are all known.

Setting $\partial R/\partial b = 0$ leads to a second linear equation:

$$m\sum x_{i,\text{obs}} + b = \sum y_{i,\text{obs}}$$

Thus, we have two equations and two unknowns and can solve for m and b directly.

Data we measure will contain experimental errors. How do we account for them? The errors can be modeled in many ways. If we assume that they follow a Gaussian distribution, then it can be shown that the best approach is to minimize

$$\sum \left(\frac{\Delta y_i}{\sigma_i}\right)^2$$

rather than

$$\sum (\Delta y_i)^2.$$

Here σ_i is the standard error of the ith measurement. In other words, we are doing a weighted minimization, in which the individual data points are up- or down-weighted according to their reliability.

The Gaussian model, although easy to understand, is not the most appropriate model for crystallographic data. In recent years more sophisticated statistical methods for error estimation have been introduced and have greatly improved crystallographic computing. Modern refinement programs do not simply minimize the weighted square of the difference, as described above (and shown in equation 7.1). Instead, they apply so-called maximum likelihood statistical methods to arrive at the model most consistent with the data. The error propagation models required to estimate probabilities can be complicated, but the general idea remains the same—data points are still weighted on the basis of their reliability. For the sake of simplicity, we will only describe the least-squares method in the following discussion.

For the straight line discussed earlier, the y_{calc} values depend on two parameters, the slope m and intercept b. In the crystallographic case, the calculated structure factor F_{calc} is a function of the model parameters $p_1, p_2, \ldots, p_N$, which are primarily the x, y, and z coordinates of the different atoms. The set of parameters p_i is often abbreviated as the array or vector $\boldsymbol{p}$.

By analogy with what we did with the straight line, we can define a residual R as the weighted square of the difference between observed and calculated structure factors, summed over all reflections:

$$R = \sum_h \left(\frac{F(h)_{obs} - F(h)_{calc}}{\sigma_h} \right)^2 \tag{7.1}$$

Note that the residual is based on the observable amplitudes $|F|$ and thus does not require knowledge of phases. The best estimates of the model parameters are those that give the minimum value for R. We find those optimal parameters by differentiating R with respect to each parameter and setting the derivatives to zero, just as we did earlier to find m and b.

$$\frac{\partial R}{\partial p_1} = 2 \sum_h w_h (F_{obs} - F_{calc}) \frac{\partial F_{calc}}{\partial p_1} = 0$$

$$\frac{\partial R}{\partial p_2} = 2 \sum_h w_h (F_{obs} - F_{calc}) \frac{\partial F_{calc}}{\partial p_2} = 0$$

$$\vdots$$

$$\frac{\partial R}{\partial p_N} = 2 \sum_h w_h (F_{obs} - F_{calc}) \frac{\partial F_{calc}}{\partial p_N} = 0$$

where w_h is a weighting term (for example, $1/\sigma_h^2$). If the model has N parameters, we will obtain a set of N normal equations, one for each parameter p_i. As long as N is less than the number of data points (reflections), we can hope to solve this set of simultaneous equations and determine the best values for p. However, if N exceeds the number of reflections, the system is said to be under-determined, and no unique solution for p can be found. In practice, it is desir-able to have a heavily overdetermined system, with the number of data points being $\gg N$. Overdetermination ensures that the estimation of p is robust.

As we saw in the earlier example, a linear function yields a simple set of linear normal equations that are easily solved. The structure factor equation is *not* linear, however. Recall Equation (3.14):

$$F_{calc} = \sum f_j e^{2\pi i(hx_j + ky_j + lz_j)} e^{(-B\sin^2\theta/\lambda^2)}$$

This is a key difference between fitting a straight line and crystallographic refine-ment. Differentiating the structure factor equation gives rise to a very complex and nonlinear set of normal equations that cannot be solved directly.

To overcome this problem, we must resort to approximations. Many possibil-ities exist. The simplest calculations are obtained when using highly simplistic approximations, but these are too unrealistic to be useful. More realistic approx-imations give rise to more difficult calculations. A reasonable compromise uses a Taylor series to obtain a linear approximation to the structure factor, which we can plug into the expression for R. We then calculate derivatives and set them to zero, generating a set of normal equations.

It is at this point that the need for a preliminary model becomes clear. A Taylor series expansion is only valid in a limited range around the original set of values. In our case, these original values are the set of coordinates from the initial model. Solving the normal equations allows us to calculate *corrections* Δp_i to be applied to each parameter (think of it as tweaking the positions of the atoms). Because we are using an approximation to the structure factor, the corrections we calculate are not perfect. Hence, we must carry out a series of successive approx-imations. In each refinement cycle:

1. We apply the corrections, calculating $p_{i,new} = p_i + \Delta p_i$;
2. Minimize again using $p_{i,new}$, and calculate $\Delta p_{i,new}$; and
3. Return to step 1.

The refinement converges when R no longer deceases from cycle to cycle.

Our systems of normal equations are really big. For example, a 35-kDa protein will contain about 300 residues, or about 2,400 nonhydrogen atoms. Each atom has x, y, and z coordinates, which gives a model with $3 \times 2,400 = 7,200$ parameters. This means that the set of normal equations will contain $7,200 \times 7,200 \approx 5 \times 10^7$ coefficients. In the language of linear algebra, to solve the equations we would have to invert a normal matrix containing 5×10^7 elements. Inverting such a large matrix is computationally intractable. To get around this problem, we take additional shortcuts that allow our computers to solve the equations quickly; even with these shortcuts, however, the calculations are impressive in their scale. Until quite recently, refining even small to moderate-sized proteins required weeks of CPU time. Now, refinements require hours to days.

An important question is whether we have sufficient data to support the refinement of complex atomic models containing thousands of parameters. Suppose the 35-kDa example from the previous paragraph crystallizes in the space group $P222$, with unit cell lengths $a = 60$ Å, $b = 70$ Å, and $c = 80$ Å. A 3 Å data set from this crystal will contain only about 6,500 unique reflections, and a 2 Å dataset will contain about 22,000 unique reflections. Therefore, the refinement is underdetermined at 3 Å, and only threefold overdetermined at 2 Å. In contrast, many small molecule structures are refined with data/parameter ratios of 20 or 30. Therefore, it appears that *we simply don't have enough diffraction data to refine a typical protein structure.*

We can surmount this problem by drawing on the tremendous body of knowledge that is available about the chemical structures of protein molecules (bond lengths, angles, and so on). This information is derived from high-resolution crystal structures of amino acids and small peptides, as well as from spectroscopy and theoretical studies. We can incorporate this knowledge into our refinement in the form of appropriately weighted stereochemical restraints, which act like additional data. Thus, the residual that is minimized would contain many additional summations, in addition to the structure factor terms shown in equation (7.1). An example might be

$$\sum_{i=1}^{\#\,\text{bonds}} \left(\frac{b_{i,\text{obs}} - b_{i,\text{std}}}{\sigma_{\text{bond},i}} \right)^2$$

Here the b_i are the lengths of chemical bonds. The "observed" values are those found in the model, and the standard values are readily available in the literature.

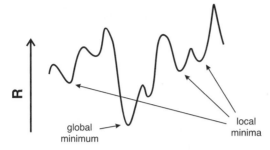

R

global
minimum

local
minima

Figure 7.2. A one-dimensional example illustrating the notion of local minima in refinement. In this figure the horizontal axis represents some parameter that is changing during the refinement (an atomic position, for example). The y axis shows the R value and is hence a measure of the how well the model agrees with the data. When a model becomes trapped in a local minimum, it can be difficult for the refinement program to extricate it and proceed to the global minimum, since moving the parameter by small amounts in either direction worsens the R value.

The σ is a weight that determines how strongly this particular restraint contributes to the refinement. Keeping this sum small ensures that the bond lengths in the model stay close to the known values $b_{i,\text{std}}$. The $b_{i,\text{std}}$ values play the same role in the normal equations as the F_{obs} values do and therefore act as additional observations. This increase in the number of observations makes the refinement effectively overdetermined.

   The method outlined in this chapter assumes that the parameters in the starting model are reasonably close to the correct values (in which case we say the model lies "within the radius of convergence"). This is not always possible, and a poor initial model can make it difficult for the refinement to reach the correct answer. Think of the minimization function as a highly convoluted multidimensional surface; there is one correct answer (the global minimum), but many local minima may lie between your current position and the global minimum. A one-dimensional example of a minimization function is schematized in Figure 7.2.

   It is difficult for the refinement program to traverse local minima. A good example would be a flipped peptide bond that was built into a model with the omega angle = 180° when the correct value is actually $\omega = 0°$. The model with ω = 180° isn't very good, but it is better than a model with $\omega = 90°$, so the refinement program will never move omega sufficiently far to discover its correct value. When a refinement becomes "stuck" before reaching an appropriate R value, the model must be inspected and carefully compared with the electron density map. Frequently at this point the crystallographer can detect problems such as the flipped peptide bond mentioned above. After any such gross problems are corrected, the refinement cycles can be resumed.

Another way to circumvent the problem of local minima is *simulated annealing*. In this approach, the thermal motions of the molecule are simulated in the computer. The atoms are allowed to move subject to Newton's equations of motion; they are restrained to obey simple chemical laws so that bond lengths and angles remain reasonable. The idea behind this is that the thermal motion will allow the molecule to sample many different conformations and so escape from any local minima in which it may have been trapped. The geometric restraints are applied by writing an equation for the potential energy of the molecule as a function of the coordinates. It contains, for example, terms for bond stretching, hydrogen bonding, and charge interactions. Many of terms in this equation look similar to the restraint terms in least-squares refinement. In addition, an "X-ray force"—arising from the familiar term $(F_o - F_c)^2$—is applied that biases the atomic motions toward those that reduce R. Because this bias is weak, the molecule can make motions for which R increases temporarily and that can allow it to surmount some local minima. The refinement starts at high temperature, with the molecule making large motions. The motions are then slowly damped by reducing the temperature until the molecule settles down into a final structure.

The metric that is most commonly used to monitor a refinement is the crystallographic R value:

$$R = \frac{\sum_{hkl} (|F_{obs} - F_{calc}|)}{\sum_{hkl} F_{obs}} \qquad (7.2)$$

As the refinement progresses and the model improves, R decreases. Note that this R value is not exactly the same as the R values we defined earlier (equation 7.1), which are the minimization targets for the refinement. However, the crystallographic R value is obviously very similar to the target function being minimized. The R given in Equation 7.2 is the one usually quoted when describing the structure.

Powerful refinement methods, such as simulated annealing, can sometimes drive down the R value by unrealistically distorting the model rather than by improving it. Restraints control this problem but do not always eliminate it. The *free R value* was introduced to address the problem of overfitting. At the outset of the refinement, a random subset of the reflections is set aside. This is referred to as the *test set* and typically corresponds to about 5% of the total. The balance of the reflections is known as the *working set*. During refinement, the minimization uses only the reflections in the working set. The R value for the reflections in the working set will almost always decrease during refinement, but if the model is truly improving, then the R value for the test set (known as R_{free}) should also decrease. Typically $(R_{free} - R) < 0.1$.

7.3 Summary

- Crystallographic structure determinations lead to models showing the spatial location of the atoms of the protein.
- Initial models can contain significant errors. Refinement improves these models by systematically adjusting the atomic positions to maximize the agreement between the observed diffraction data and data calculated from the model.
- Because proteins and other macromolecules typically do not diffract to very high resolution, the refinement of their structures suffers from a poor ratio of data to parameters, which causes the refinements to be ill conditioned. To overcome this problem, stereochemical restraints are included which ensure that parameters, such as bond lengths, angles, and so on, remain close to known values. These restraints act as additional data points, making the refinements effectively overdetermined.
- Refinement by least squares is a generalization of the familiar fitting of a straight line to a set of points. Least-squares refinement is an iterative process, the progress of which is measured by both the crystallographic R value and by how closely the model adheres to stereochemical rules. Modern refinement programs use error treatments that are more sophisticated than that associated with least-squares, but the overall approach to refinement remains similar.
- Refinement by simulated annealing can help models to escape false minima.

FURTHER READING

Least-squares analysis is covered with great clarity in *X-ray Crystal Structure Determination: A Practical Guide* by Stout and Jensen. Lyle Jensen introduced refinement in protein crystallography in the face of deep skepticism.

See also the following paper, which gives a beautiful, didactic discussion of protein refinement. Cruickshank, DWJ. Remarks about protein structure precision. *Acta Crystallogr. D* 1999;55(3):583–601.

The advantages and perils of simulated annealing and related methods are well described in Brunger AT, Adams PD. Molecular dynamics applied to X-ray structure refinement, *Acct. Chem. Res.* 2002;35(6):404–412.

A brief and readable account of the application of maximum likelihood statistical methods to crystallographic problems such as refinement is given in the following paper: McCoy AJ. Liking likelihood. *Acta Crystallogr. D* 2004;60(12):2169–2183.

Glossary

Anomalous scattering / anomalous difference When X rays are scattered by a free electron, a phase change of 180° occurs, so that the scattered radiation is out of phase with the incident radiation. Under most circumstances the electrons in an atom can be regarded as free, so that when calculating the diffraction by a crystal (structure factors) it is customary to assume this 180° phase shift. However, if the energy of the X rays is near an absorption edge of a particular atom type, this assumption is no longer valid, and the phase change for scattering from such atoms will differ from 180°. This is represented mathematically by allowing the atomic scattering factor f to be complex: $f = f_0 + f' + if''$. Here f_0 is the normal scattering factor, and f' and f'' are increments arising near the absorption edge. Anomalous scattering is manifested in two ways: the *Friedel's law* relation $F(h,k,l) = F(-h,-k,-l)$ breaks down, and $F(h,k,l)$ becomes a function of wavelength near the absorption edge. These effects allow one to determine the absolute handedness of a molecule and to determine phases if an anomalous scatterer is present in the crystal. The quantity $| F(h,k,l) - F(-h,-k,-l) |$ is called the anomalous difference.

Asymmetric unit The smallest substructure of a crystal that, when repeated by the space group operations, can generate the entire crystal. The asymmetric unit is often a single molecule but can also be an oligomer.

B factor / B value A parameter that describes the falloff of diffracted intensities with increasing $\sin\theta/\lambda$ that arises from thermal motion or disorder. The calculated structure factors are multiplied by a *temperature factor* or *Debye-Waller factor* $\exp(-B\sin^2\theta/\lambda^2)$ to account for this effect. In simple models $B = 8\pi^2 u^2$, where u^2 is the mean-square amplitude of vibration of an atom. The customary units of B are Å^2.

Bijvoet pair A pair of reflections $F+$ and $F-$, where $F+$ is either $F(h,k,l)$ or a symmetry-related reflection (i.e., a reflection whose intensity is required by the symmetry of the diffraction pattern to be equal to $F(h,k,l)$); and $F-$ is either $F(-h,-k,-l)$ or a symmetry-related reflection. $F+$ and $F-$ have equal amplitudes for normal scattering but will have unequal amplitudes in the presence of *anomalous scattering*. See also *Friedel pair*.

Bragg reflection A particular point in the diffraction pattern of a crystal where *Bragg's law* predicts that significant diffracted intensity will occur (i.e., one of the "spots" in the diffraction pattern). So named because, in the Bragg model of diffraction, the incident X-ray beam appears to be reflected by families of planes in the crystal when *Bragg's law* is satisfied. Bragg reflections are labeled by the *Miller indices h,k,* and *l*, which denote which Bragg plane gives rise to a particular reflection.

Bragg's law Arises from a schematic model of crystal diffraction in which the crystal is represented by families of parallel planes spaced d units apart. If the incoming X-ray beam makes an angle θ with the planes, Bragg's law states that for integer values of n, whenever $n\lambda = 2d\sin\theta$ the incoming X-ray beam will appear to be reflected from the family of planes and will flash out of the crystal as a beam of X rays, creating a spot or *reflection* on the detector.

Centric reflections Reflections whose phase is constrained by crystal symmetry to be either $0°$ or $180°$. For *centrosymmetric* crystals all reflections are centric. Noncentric crystals may still have zones of reflections that are centric.

Centrosymmetric In a centrosymmetric crystal, for each atom at position x,y,z there is an identical atom (a symmetric mate) at $-x, -y, -z$. In centrosymmetric crystals all reflections are *centric*. Crystals containing chiral molecules, such as L-amino acids, cannot be centrosymmetric unless the crystal contains the racemic mixture.

Chiral volume A tool used during refinement to ensure that the configuration of substituents around α-carbon atoms or other chiral centers has the correct hand. A parallelepiped is defined for each such center; the volume of this parallelepiped is given by the vector relation $v_1 \cdot (v_2 \times v_3)$, where v_1, v_2, v_3 are interatomic vectors, defined so that the volume will be positive if the center has the correct chirality. The chiral volume is restrained to be positive during refinement.

Combination of phase information For a given reflection hkl and a given source of phase information, say MAD, the probability that the phase angle α is correct is given by $P_{MAD}(\alpha[hkl])$. There are various models to compute P_{MAD} from experimental data. If there is an additional source of phase information, say MIR, then one can generate a corresponding phase probability curve $P_{MIR}(\alpha[hkl])$. The overall probability that the phase angle α is correct is given by the product of these two curves: $P_{TOT} = P_{MAD}P_{MIR}$. Blow and Crick showed that the best phase α_{BEST} is given by the centroid phase of the curve P_{TOT}.

Debye-Waller factor The *temperature factor* as applied to diffraction intensities: $\exp(-B[\sin\theta/\lambda]^2)$. See *B factor*.

Density modification A method to improve electron density maps and to refine phases. Density modification restores to the electron density map known characteristics that have been distorted by phase errors. For example, the bulk solvent regions between molecules should have a known, constant electron density, but this is not the case in actual maps. Solvent flattening is a procedure in which the solvent regions of the unit cell are set to the correct, constant value. The modified density is then Fourier transformed to yield new structure factors $F_{calc}\exp(i\alpha_{calc})$. Hybrid structure factors $F_{obs}\exp(i\alpha_{calc})$ are then used to generate a new electron density map. The solvent regions in this map will be flatter than before, but still not flat. The process is repeated until the density in the solvent region no longer changes. In general, the final maps are improved in all regions, not only in the solvent, so that poor density within the molecule may become clearer. *Histogram matching* and *noncrystallographic symmetry* averaging are other examples of density modification.

Difference maps Difference electron density maps (difference Fouriers) are used to reveal unmodeled features in the structure. These maps are calculated from Fourier syntheses by using coefficients $(F_{obs} - F_{calc})\exp(i\alpha_{calc})$. Positive peaks in difference Fouriers represent features present in the structure that are not accounted for by the model. Negative peaks represent features in the model that are not present in the structure. Adjacent positive and negative peaks often suggest that an element has to be shifted in the model. Because the coefficients are approximations, peaks in these maps come up at half their true weight. Maps created using coefficients $(2F_{obs} - F_{calc})\exp(i\alpha_{calc})$ add together the current calculated map plus twice the difference map to approximate a corrected map.

Difference Patterson maps are Fourier syntheses calculated using coefficients $(F_{PH} - F_P)^2$ where P stands for protein and PH stands for protein+heavy atom. Such maps reveal the set of vectors running between heavy atoms and are frequently used to locate such atoms. These maps contain noise since the coefficients are approximations.

Direct methods Methods used to determine crystal structures of small molecules ($<\sim100$ nonhydrogen atoms). Developed by Karle and Hauptman and by Sayre, direct methods restrict the set of possible phases by imposing physical constraints on the electron density function ρ: $\rho \geq 0$ and $\rho^2 \sim \rho$ are important examples. Direct methods are probabilistic in nature and rely for their effectiveness on having data to very high resolution. Classical direct methods are not useful for proteins, in general, since the width of the phase probability curves increases with the number of atoms in the structure and becomes essentially flat for large molecules. However, such direct methods have been very successful in determining the heavy atom or anomalous scatterer substructures within proteins. Modern direct methods employing *dual space* approaches have proven capable of determining even very large structures, so long as atomic resolution data are available.

Dispersive difference Dispersive means varying with wavelength. The MAD method makes use of differences in the structure factor amplitudes at different wavelengths. When $F(h,k,l)$ is measured at the two wavelengths $\lambda 1$ and $\lambda 2$, the quantity $|{}^{\lambda 1}F(h,k,l) - {}^{\lambda 2}F(h,k,l)|$ is called the *dispersive difference*. See *anomalous scattering*.

Dual space methods Phasing methods that alternate between *reciprocal space* (i.e., diffraction space–structure factors) and real space (the electron density map). A simplistic example of how this might work would be to alter the phase of each reflection (randomly, or according to some criterion of quality); this is in reciprocal space. A map would then be calculated with the modified phases and subjected to density modification (real space). The phases resulting from density modification would be modified again, and the process repeated until convergence. Dual space methods, as implemented in programs such as SnB and SHELX, have been tremendously successful at extending the applicability of direct methods to macromolecules.

Electron density (function or map) X rays are scattered by electrons, and the strength of the scattering by a tiny volume is proportional to the number of electrons in that volume or to the electron density. The image created in crystallography is the electron density function, represented graphically as an electron density map.

Error analysis In crystallography this term usually refers to analyzing errors arising during estimation of the phases. In MIR, for example, it is rare for the phase circles to meet at a single point, and no unique value for the protein phase α_P emerges from the calculation. Rather, we say that multiple values of α_P are consistent with the observations and their associated errors, each with a different probability of being correct. The result of an error analysis is a set of curves $P(\alpha)$, one for each reflection, that give the probability that a particular phase α is the correct value of α_P. See *combination of phase information*.

Eulerian angles A set of three angles or rotations used to rotate an object in three dimensions. In one convention the first rotation (θ_1) is applied to the x axis, the second (θ_2) is applied to the moved y axis, and the third (θ_3) is applied to the twice-moved z axis. For crystallographic calculations, these rotations are applied to a molecule by using a rotation matrix.

Ewald's sphere A geometric construct designed to aid in the visualization of *Bragg reflections*. A sphere is drawn centered on the crystal, with a radius of $1/\lambda$, where λ = the wavelength of the incident X-ray beam. The point where the beam exits the sphere is defined as the origin of the *reciprocal lattice*. The axes of this lattice correspond to the *Miller indices* h, k, and l, so that every lattice point corresponds to a particular Bragg reflection. Rotating the crystal rotates the reciprocal lattice. When a particular reciprocal lattice point lies on the surface of the Ewald sphere, *Bragg's law* is satisfied for that reflection. The direction of the diffracted beam is obtained by drawing a ray from the center of the sphere through the position on the surface where the reciprocal lattice point lies.

Figure of merit (FOM) The FOM, often denoted m, is a statistic indicating the reliability of the experimental estimate for a particular phase. It is a measure of the breadth of the phase probability curve $P(\alpha)$ for that reflection. The smaller m, the broader the curve. It is roughly true that $m = \cos \Delta\alpha$, where $\Delta\alpha$ is the estimated phase error; hence $m = 0.5$ corresponds to an estimated error of 60°. The mean value of m, $<m>$, is often quoted as a statistic describing the quality of the phase determination.

Fractional coordinates If x, y, z are the coordinates of a point in Å units along the unit axes a, b, c, then the fractional coordinates of the same point are $x/|a|$, $y/|b|$, $z/|c|$.

Free R value See R_{free}

Friedel pair The pair of centrosymmetrically related reflections $F(h,k,l)$ and $F(-h,-k,-l)$.

Friedel's law The relation $|F(h,k,l)| = |F(-h,-k,-l)|$, strictly valid only for normal scattering. See *anomalous scattering*.

Hanging drop A crystal-growing method. A small reservoir (usually a well in a 24- or 96-well plate) is filled with a crystallization solution. A small volume (~ 1 μl) of the well solution is pipetted onto a coverslip and mixed with an equal volume of protein solution. The cover slip is then turned drop-side down and suspended over the well, and the reservoir is sealed with grease or tape. The protein drop becomes concentrated as water vapor moves from the drop to the well to equalize osmotic strength. From time to time this concentration process produces a crystal in the drop.

Harker section A special section in the *Patterson function* in which vectors between symmetry-related atoms appear. In the space group $P2$, for example, for each atom at position x,y,z there is another at position $-x,y,-z$. The interatomic vectors between these are *2x, 0, 2z* and *− 2x, 0, −2z*. The section $y = 0$ in the Patterson function is a Harker section for this space group and contains all the vectors between atoms related by the twofold rotation. Harker sections are useful in determining the positions of heavy atoms.

Heavy atom refinement A method of improving the estimate of the positional and other parameters associated with heavy atoms to improve the calculation of phases. In MIR one has measured values for $F_{PH\text{-}obs}$. One also has values for $F_{PH\text{-}calc}$ = $F_P + F_H$. Using *least-squares refinement,* one can minimize the residual $\Sigma(F_{PH\text{-}obs} - F_{PH\text{-}calc})^2$ with respect to the heavy atom positions and other parameters used in calculating F_H. The process is cyclical. After improving the calculation of F_H one redoes the phase calculation for F_P, yielding a new estimate of $F_{PH\text{-}calc}$ to use in the next refinement cycle.

Heavy atoms Atoms containing many electrons. Various heavy atom compounds are used to derivatize protein molecules in the *isomorphous replacement* method. Mercury compounds, for example, react with free –SH groups.

Histogram matching A method of *density modification.* The values of the electron density sampled at regular intervals (grid points) throughout the three-dimensional map form a probability distribution that is represented as a density histogram. For error-free structures this histogram has a characteristic profile. The histogram-matching method takes the density histogram calculated from an initial set of phases and modifies it so that it takes the form of an expected density histogram.

Huygens principle A principle by which the diffraction pattern of an object can be calculated. It assumes that each volume element in the object is a source of scattered spherical waves, whose amplitude is proportional to the scattering power of that volume element. The object's diffraction pattern as seen by a distant observer is the sum of all the spherical waves emanating from it.

Insertion device A device that allows X rays to be generated in the straight segments of a *synchrotron* ring. Arrays of magnets are inserted into the path of the electron beam to force the beam into an oscillating path, thereby inducing the production of X rays.

Intensity The observed strength of an X-ray reflection, proportional to the number of X-ray photons striking the detector during the measurement period. The intensity is proportional to the square of the structure factor amplitude: $I(h) \propto |F(h)|^2$.

Isomorphous replacement A method of phase determination for macromolecular crystals. Structure factor amplitudes $F_P(h)$ are measured from a native protein crystal. Crystals derivatized with a heavy atom compound are then prepared and screened for isomorphism and quality. If a suitable crystal is found, structure factor amplitudes $F_{PH}(h)$ are measured from it. Using direct or Patterson methods the coordinates of the heavy atoms are determined, and amplitudes and phases of the heavy atom structure factors $F_H(h)$ are calculated. The structure factors for the native and derivative are related by the equation $F_P(h)\exp(i\alpha_P) + F_H(h)\exp(i\alpha_H) = F_{PH}(h)\exp(i\alpha_{PH})$.

There is one equation for each reflection. We would like to solve each of these equations for α_P. We already know the four quantities F_P, F_{PH}, F_H, and α_H. It turns out that the equation above is satisfied by two distinct values of α_P. To eliminate this ambiguity a second heavy atom derivative must be prepared, leading to a second set of equations like the one above. In most cases the two equations will have only one solution for α_P in common.

Lattman angles Variants of *Eulerian angles* that produce undistorted rotation function maps, in which different peaks can be accurately compared. In the Euler angle system, when θ_2 is small the rotations θ_1 and θ_3 occur about axes that are nearly parallel, and are strongly correlated. This correlation can be greatly reduced by using the modified angles: $\theta_+ = (\theta_1 + \theta_3)/2; \theta_2 = \theta_2; \theta_- = (\theta_1 - \theta_3)/2$.

Laue method A method to record a diffraction pattern very rapidly by using an X-ray beam containing a broad spectrum of wavelengths. For beams containing all wavelengths between λ_{min} and λ_{max}, the *Ewald's sphere* construction involves two spheres of radii $1/\lambda_{min}$ and $1/\lambda_{max}$. Any reciprocal lattice point lying between these two spherical shells will satisfy *Bragg's law* for one of the intermediate wavelengths and will give rise to a reflection on the detector. Laue experiments must be carried out at *synchrotrons*, which provide intense white radiation. In favorable cases a significant fraction of the reflections can be measured for a fixed crystal setting, allowing nearly complete data sets to be recorded in a fraction of a second.

Least-squares refinement A method of improving the accuracy of the atomic model in an X-ray crystal structure. Least-squares refinement minimizes the residual $R = \sum_{h}(F_{obs}(h) - F_{calc}(h))^2$. One can linearize F_{calc} in a power series, differentiate F_{calc} with respect to model parameters (like atomic coordinates), and then set the derivatives to zero. This gives a set of *normal equations* that contain terms such as the following:

$$\left(\frac{\partial F_{calc}}{\partial u_j}\right)\Delta u_j$$

where Δu_j is a shift in the x or y or z coordinate of an atom.

The math of minimizing R is now identical with that of finding the best line through a set of points. One recovers shifts Δu_j for all the coordinates, and these are added to the existing coordinates to provide updated, improved positions. Because the linearization of F_{calc} is quite inaccurate, the calculated shifts Δu_j are not completely accurate, and one has to repeat the refinement protocol many times, each time using the updated coordinates for the expansion of F_{calc}, until the process converges and R is minimized.

MAD An acronym for *multiple wavelength anomalous diffraction*

Miller indices Integer indices *h, k,* and *l*, used to identify families of Bragg planes. The (h,k,l) family of planes divides the unit cell vector *a* into *h* equal parts, the unit cell vector *b* into *k* equal parts, and the unit cell vector *c* into *l* equal parts.

MIR An acronym for multiple isomorphous replacement. See *isomorphous replacement.*

MIRAS A method of phase determination combining multiple isomorphous replacement and anomalous scattering.

Mirror plane A symmetry element in a crystal that relates two structures by a mirror operation. For example, if the *z* axis is normal to the mirror plane, then the equivalent positions generated by the mirror are x, y, z and $x, y, -z.$

Molecular replacement A method of phase determination that does not require heavy atoms or anomalous scattering. It is useful only when the structure of the molecule of interest is similar to that of a reference molecule of known structure. Molecular replacement works by creating a model of the unit cell of the crystal of interest in which the reference molecule appears in the same position and orientation as the molecule of interest. This model can then be refined. Placing a molecule of known structure in a unit cell requires six numbers: three coordinates for the center of mass and three angles to specify the orientation. Molecular replacement finds these six numbers using two successive three-dimensional searches. First the *rotation function* determines the molecule's orientation in the unit cell; then the *translation function* fixes its position.

Multiple isomorphous replacement See *isomorphous replacement.*

Multiple wavelength anomalous diffraction (MAD) A method of phase determination using anomalous scatterers that are present in the crystal. The current favorite anomalous scatterer is selenium (Se), which is biosynthetically introduced into the protein in the form of selenomethionine. Data from a crystal containing an anomalous scatterer are usually collected at at least three wavelengths: one remote from the absorption edge and two others at points within the absorption region precisely chosen to maximize signal. Within each of the three data sets *Friedel's law* is violated, so that $F(\boldsymbol{h}) \neq F(-\boldsymbol{h})$. In addition, the values of $F(\boldsymbol{h})$

vary among these three experiments because of the variation in f' and f'' with wavelength (see *anomalous scattering*). In a MAD experiment the positions of the anomalous scattering atoms are first located by direct methods or from an anomalous difference Patterson map. Then, using methods analogous to those in MIR, phases can be found. Heuristically, the three data sets in MAD are analogous to the native and two derivative data sets collected in MIR. The signal in this method is very weak and careful data collection is required.

Noncentric reflections Reflections whose phases are not constrained to be either $0°$ or $180°$. See *centric reflections.* Most reflections from protein crystals are noncentric.

Noncrystallographic symmetry Symmetry that occurs within the asymmetric unit, but which is not displayed by the crystal as a whole. A good example can be seen in crystals of icosahedral virus particles. These viruses possess fivefold axes of symmetry, but fivefold symmetry is not allowed in crystal lattices. Hence fivefold symmetry exists locally within the envelope of each virus, but not globally throughout the crystal.

Normal equations and matrix In *least-squares refinement* there is one equation for each independent reflection. A typical equation for reflection j can be schematized as

$$F_{obs}(\mathbf{h}_j) - F_{calc}(\mathbf{h}_j) = \Delta F = \cdots + \frac{\partial F_{calc}}{\partial u_m} \Delta u_m + \cdots$$

where Δu_m is the desired shift in an atomic coordinate. In general, the number of reflections N_r is larger than the number of coordinate parameters N_p. The family of equations can be summarized schematically as

$$\begin{pmatrix} \Delta F^1 \\ \vdots \\ \Delta F^{Nr} \end{pmatrix} = \begin{pmatrix} \dfrac{\partial F_c^1}{\partial u_1} \Delta u_1 + \cdots + \dfrac{\partial F_c^1}{\partial u_{Np}} \Delta u_{Np} \\ \vdots \qquad\qquad \vdots \\ \dfrac{\partial F_c^{Nr}}{\partial u_1} \Delta u_1 + \cdots + \dfrac{\partial F_c^{Nr}}{\partial u_{Np}} \Delta u_{Np} \end{pmatrix}$$

If we collect all the ΔF's and all the Δu's as column vectors then this family of equations can be written in matrix form as $\Delta F = M \cdot \Delta u$ where M is a nonsquare matrix containing the partial derivatives shown in the brackets above. Families of linear equations are much more readily handled when the matrix is square. If we multiply both sides of the above equation by M^T, the transpose of M, we get

$$M^T \cdot \Delta F = M^T \cdot M \cdot \Delta u = N \Delta u$$

This modified set of equations is called the normal equations, and the matrix N is called the normal matrix. The elements of N contain products of pairs of the partial derivatives shown above.

OMIT map An electron density map calculated to reduce the bias introduced by phases calculated from the model. OMIT maps are often made to examine a troublesome region in the normal map. OMIT maps are generated by using observed amplitudes and phases calculated from that part of the model that is outside the region being examined. In other words, the region of interest is omitted from the phase calculation. In composite OMIT maps the unit cell is divided into blocks, and an OMIT map is calculated separately for each block. The block OMIT maps are then plotted together to give a composite map of the whole unit cell.

Oscillation camera A data collection device that oscillates a crystal back and forth through a small angle while data are being collected.

Patterson function A Fourier synthesis of the form

$$P(\boldsymbol{u}) = \frac{1}{V} \sum_{\boldsymbol{h}} I(\boldsymbol{h})\cos(2\pi\boldsymbol{h}\cdot\boldsymbol{u})$$

The Patterson is defined in the same unit cell as the electron density function. However, the phases in the Patterson function are all set to zero. The Patterson coefficient uses the intensity I instead of the structure factor amplitude F. Peaks in the Patterson function correspond to interatomic vectors in the structure. A peak will occur at position $\boldsymbol{u}$ if there are two atoms with positions $\boldsymbol{x}$ and $\boldsymbol{x}'$, such that $\boldsymbol{x} - \boldsymbol{x}' = \boldsymbol{u}$. The Patterson function is useful in determining heavy atom structures.

Phase combination See *combination of phase information.*

Phase extension A method of extending knowledge of phases to higher resolution. Frequently experimental methods of phase determination (e.g., MIR) become ineffective beyond a certain resolution, even though the diffraction pattern extends farther. For example, lack of isomorphism or poor diffraction from derivative crystals might cause MIR phases to be useful only to 3 Å, even though the native crystals diffract to 2 Å. Methods such as density modification and symmetry averaging can be used to extend the phases to the limit of diffraction.

Phase problem To create an electron density map we need both the amplitude and the phase angle of the structure factor $\boldsymbol{F}$. However, we can obtain only the amplitude experimentally; the phase cannot be directly measured.

Phase refinement The methods used in *phase extension* can also be used to improve estimated phases.

Phasing power A statistic used to show the size of the heavy atom or anomalous scattering contribution compared with the errors in measurement.

The isomorphous phasing power is:

$$\left[\frac{\displaystyle\sum_{hkl} F^2_{H-calc}}{\displaystyle\sum_{hkl} \Delta F^2_{PH}} \right]^{1/2} \qquad \text{where} \quad \sum_{hkl} \Delta F^2_{PH} = \sum_{hkl} \{ F_{PH-obs} - F_{PH-calc} \}^2$$

For anomalous scatterers two phasing power statistics are used. The anomalous phasing power is:

$$\left[\frac{\displaystyle\sum_{hkl} F^2_{H-imag}}{\displaystyle\sum_{hkl} \{ \Delta F^{+/-}_{PH-obs} - \Delta F^{+/-}_{PH-calc} \}^2} \right]^{1/2}$$

where $\Delta F^{+/-}$ is the structure factor amplitude difference between Bijvoet pairs, and $F_{H\text{-}imag}$ is the imaginary part of the calculated structure factor contribution from the anomalously scattering atoms.

The dispersive phasing power is

$$\left[\frac{\displaystyle\sum_{hkl} F^2_{H-real}}{\displaystyle\sum_{hkl} \{ \Delta F^{+/-}_{PH-obs} - \Delta F^{+/-}_{PH-calc} \}^2} \right]^{1/2}$$

where $\Delta F^{+/-}$ is the structure factor amplitude difference between Bijvoet pairs, and $F_{H\text{-}real}$ is the real part of the calculated structure factor contribution from the anomalously scattering atoms.

R *value or* R *index* A statistic that measures agreement between observed and calculated structure factors. If F_{obs} and F_{calc} are the observed and calculated structure factor amplitudes then

$$R = \frac{\displaystyle\sum_{h} |F_{obs} - F_{calc}|}{\displaystyle\sum_{h} F_{obs}}$$

R is sometimes also defined for intensities instead of amplitudes. For protein crystals a value of R less than .20 is considered quite good.

Radiation damage Degradation of a crystal from the energy deposited in it by the X-ray beam. Cooling crystals to very low temperatures ($\sim$100K) reduces radiation damage.

Ramachandran plot A plot showing the allowed values of the polypeptide backbone torsion angles ϕ and ψ. Certain regions of this plot are forbidden because they result in van der Waals crashes. In a well-refined structure almost all amino acids should have ϕ- and ψ-values in the allowed ranges.

R_{Cullis} *(ano)* A statistic that indicates the agreement between observed and calculated *Bijvoet pair* differences.

$$R_{Cullis}(ano) = \frac{\sum_h \| \Delta F^{\pm}_{PH-obs} | - | \Delta F^{\pm}_{PH-calc} \|}{\sum_h | \Delta F^{\pm}_{PH-obs} |}$$

$\Delta F^{\pm}_{PH\text{-}obs}$ is the amplitude of the structure factor difference between Bijvoet pair members;

$$\Delta F^{\pm}_{PH-calc} = 2 \frac{f''}{f'} F_H \sin(\alpha_{PH} - \alpha_H).$$

R_{Cullis} *(iso)* A statistic defined for centric reflections only that indicates the quality of the phases determined for these reflections.

$$R_{Cullis}(iso) = \frac{\sum_h \| F_{PH} \pm F_P | - F_{H-calc} |}{\sum_h | F_{PH} \pm F_P |}$$

Here the $+$ or $-$ signs are chosen for each reflection so that the structure equation $F_P + F_H = F_{PH}$ is most nearly satisfied.

R_{Cullis} (λ) A statistic that indicates how well dispersive effects in anomalous scattering are accounted for by the model.

$$R_{Cullis}(\lambda) = \frac{\sum_h \| F_P(\lambda_j) - F_P(\lambda_0) | - F_{calc}(\lambda_j) |}{\sum_h | F_P(\lambda_j) - F_P(\lambda_0) |}$$

where

$$F_{calc}(\lambda_j) = \sum_{\substack{anomalous \\ scatterers}} (f'(\lambda_j) + if''(\lambda_j)) \exp(2\pi i \mathbf{h} \mathbf{x}_k)$$

Here $F_P(\lambda_j)$ and $F_P(\lambda_0)$ are structure factors at an anomalous and remote wavelength, respectively.

Reciprocal lattice An imaginary geometric construct designed to aid in visualizing diffraction. A lattice is defined in *reciprocal space* in which each *Bragg reflection hkl* from a crystal is represented by a lattice point. The unit cell vectors of the reciprocal lattice are conventionally called $\mathbf{a}^*$, $\mathbf{b}^*$, $\mathbf{c}^*$. In an orthogonal lattice $\mathbf{a}^*$

is defined as being parallel to a, with $|a^*| = 1/|a|$. b^* and c^* are defined similarly; hence, the name reciprocal lattice. (For nonorthogonal lattices the relationships between the real and reciprocal space vectors are slightly more complex).

Reciprocal space A space or coordinate system in which the three-dimensional diffraction pattern of an object is defined: so-called because small distances in direct space correspond to large distances in reciprocal space, and vice versa. Reciprocal space is the space in which the Fourier transform of the object is expressed. When the object is moved, reciprocal space moves with it. For any one orientation of the object only a subset of points in reciprocal space can be seen in a diffraction experiment. See *Ewald's sphere.*

Refinement A mathematical method for systematically improving the values of parameters, such as atomic positions, known approximately from an initial model. See *least-squares refinement* and *simulated annealing.*

Reflection See *Bragg reflection*

Resolution A descriptor of an X-ray data set, specifying the smallest value of the Bragg spacing d for which data are present (confusingly, in crystallographic parlance, "high resolution" corresponds to small values of d). Equivalently, the period of the shortest wavelength Fourier components used in synthesizing the electron density function.

Restraints, Restrained refinement Mathematical terms used in refinement that restrict the model parameters to lie within reasonable ranges. For example, if b_0 is the known average length of a certain type of chemical bond, and if the b_j values are the lengths of the various bonds of this type in a model, then including a term $\Sigma(b_j - b_0)^2$ in *least-squares* minimization will tend to keep the individual values of b_j near the canonical b_0 value. Without restraints, parameters can drift to unreasonable values during refinement, especially when the ratio of the number of X-ray reflections to the number of model parameters is too low. Restraints can be thought of as additional data points.

R_{free} A variant of the R index that is useful in detecting biases introduced by refinement.

$$R_{free} = \frac{\sum_{\text{subset of } h} |F_{obs} - F_{calc}|}{\sum_{\text{subset of } h} F_{obs}}$$

Here the subset of h represents a randomly chosen set of reflections, comprising 5–10% of the whole set, that are omitted from the refinement process. In the ideal case the refined model would predict F_{calc} in the omitted and refinement sets equally well: R_{free} would equal R. In a case of extreme overrefinement, in which much noise was fit, the refined model would predict F_{calc} for the omitted set much more poorly than for the refinement set. In practice one hopes to see $(R_{free} - R) < 0.05–0.1$.

Rigid body refinement Refinement in which only the position and orientation of a molecule or domain are varied.

$\mathbf{R}_{Kraut}$ *(ano)* A statistic used in anomalous scattering to indicate how well the model accounts for the Bijvoet differences.

$$R_{Kraut}(ano) = \frac{\sum_h | F_{PH-obs}^+ - F_{PH-calc}^+ | + | F_{PH-obs}^- - F_{PH-calc}^- |}{\sum_h |F_{PH-obs}^+ + F_{PH-obs}^-|}$$

$\mathbf{R}_{Kraut}$ *(iso)* A statistic used in isomorphous replacement to evaluate the refinement of heavy atom parameters.

$$R_{Kraut}(iso) = \frac{\sum_h | F_{PH} - | F_P + F_{H-calc} \|}{\sum_h |F_{PH}|}$$

$\mathbf{R}_{merge}$ A statistic used to indicate the quality of X-ray data. R_{merge} reflects the internal consistency of a data set: Do multiple observations of the same quantity (or quantities related by crystal symmetry) agree? If there are N observations of a given reflection, then

$$R_{merge} = \frac{\sum_h \sum_{j=1}^{N} | F_M(h) - F_j(h) |}{N \sum_h F_M(h)}$$

where F_M is the mean value of the structure factor amplitude, and the F_j are the individual measurements of the same structure factor.

Rossmann and Blow Michael Rossmann and David Blow introduced the *rotation function* into protein crystallography.

Rotation camera A data collection device that rotates a crystal through a small angle while data are being collected.

Rotation function A rotational search procedure that reveals noncrystallographic symmetry within a crystal or that determines the orientation of a known molecule or fragment in a target crystal (see *molecular replacement*.) The rotation function seeks large overlap between two *Patterson functions* as a function of their relative orientation. Equivalently, comparison of diffraction patterns can be used.

Scale factor A factor by which one multiplies calculated structure factors F_{calc} to scale them to the observed F values. Similarly for intensities.

Selenomethionine A homolog of the amino acid methionine containing a selenium atom in place of the native sulfur.

Self-rotation function A *rotation function* comparing a *Patterson function* with itself. Used to reveal *noncrystallographic symmetry*.

Self-vector A Patterson vector running between atoms in a single molecule.

Shake-and-Bake (SnB) A *direct-methods* algorithm for phase determination. SnB combines density modification in direct space (shaking) with refinement of *triplet phases* in reciprocal space (baking). SnB has been successful in solving the structures of sets of anomalous scatterers containing many atoms. It has also been used in a few cases for small proteins, when very-high-resolution diffraction is available.

Sigma A (σ_A) map A more sophisticated version of *Sim weighting*, in which errors in the known part of the structure are taken into account. So named because the factor sigma A (see *sigma A plot*) occurs in the weights.

Sigma A (σ_A) plot A graphical method of estimating the coordinate error in a refined model. Sigma A (σ_A) is a statistic that is proportional to the fraction of the structure actually included in the model, and also to a temperature factor-like quantity. Plots of $\ln(\sigma_A)$ vs. $\sin^2 \theta / \lambda^2$ have linear segments whose slope is related to the mean square coordinate error.

Sim weighting A weighting scheme to improve interpretability of electron density syntheses—for example *omit maps*—in which the phases are obtained from a partial structure. The method is based on a detailed error analysis.

Simulated annealing A method of structure refinement in which motions of protein atoms at high temperature are simulated in the computer. The motions are governed by Newton's laws, but are restrained by requiring that the R index between the moving molecule and the observed data stay small. As the molecule cools (is annealed) in the computer, it can settle down into new configurations that eliminate errors made in building the model. In other words, the model's thermal motion allows it to surmount some local minima. See the text for a fuller description.

SIR Single isomorphous replacement. Use of a single heavy atom derivative to determine phases.

SIRAS Single isomorphous replacement + anomalous scattering. Use of a single heavy atom derivative, such as mercury, plus its anomalous scattering to determine phases.

Space group A set of symmetry operations that defines the symmetry of a crystal; there are 230 possible space groups. Allowed symmetry operations include rotations (e.g., twofold axis), inversion operations (e.g., mirror plane), and translations. Combinations of such elements are also possible; a rotation plus a translation gives a screw axis. Only the 65 space groups that do not contain an inversion operation are possible for biological macromolecules (because they contain only one stereoisomer).

Spherical polar angles A scheme for representing rotations based on the idea that any arbitrary rotation can be accomplished by a spin about a single, properly chosen axis. In a common convention the angles ϕ and ψ represent the longitude and colatitude of the rotation axis, and the angle κ gives the spin around it.

Standard deviation Square root of the *variance*.

Structure factor The total scattering of radiation by a set of scatterers, often atoms. In crystals the structure factor is given by $F(S) = \sum_j f_j \exp(2\pi i x_j \cdot S)$ where f_j is the scattering factor for the jth atom at position x_j. S is known as the scattering vector and defines the direction of the scattered radiation. For a crystal the *Bragg's law* condition means that $F(S) = 0$ unless $S = h$, where h is a reciprocal lattice vector.

Symmetry averaging A form of *density modification* in which the electron density is averaged to enforce local (noncrystallographic) symmetry. For example, it is common to average the electron density of virus particles about their fivefold axes.

Symmetry operation A transformation, such as rotation, that superimposes a molecule within the crystal upon an identical copy of itself.

Symmetry-related reflections Set of two or more reflections related by symmetry within the diffraction pattern. The structure factor amplitudes of symmetry-related reflections are always equal, but the phases may display simple differences. For example, in the space group $P2$ the structure factor amplitudes $F(h,k,l)$ and $F(h,-k,l)$ are equal.

Synchrotron radiation Synchrotron radiation sources produce brilliant, highly collimated X-ray beams containing a "white" frequency spectrum. A synchrotron is a large facility in which highly energetic (Gev) electrons circulate in a polygonal ring. At the vertices of the ring, bending magnets change the direction of the electrons' path, accelerating them in the process. Synchrotron radiation is emitted by the electrons as they undergo this acceleration. Synchrotron radiation is generally focused and monochromatized to provide a beam with a tiny cross section and the desired wavelength. Ready access to synchrotron radiation has been essential for the explosive growth in *MAD* phasing.

Temperature factor See *B factor*.

Thermal motion The random displacement of atomic centers from their mean positions arising from the thermal energy of the atom. See *B factor*.

Translation function A step in the *molecular replacement* method in which the position of the reference molecule in the unit cell of the target crystal is determined.

Triplet relation A key relation in *direct methods* stating that if the reflections E_h, E_k, and E_{h-k} are all strong, then their phases are related by the expression $\phi_h \approx \phi_k + \phi_{h-k}$. This relation becomes progressively weaker as the number of reflections in the data set increases.

Undulator A type of *insertion device* producing X rays whose wavelengths are confined to particular wavelengths. It consists of a periodic magnetic dipole structure through which electrons pass. The electrons undergo oscillations and radiate. This device is related to a free-electron laser.

Unit cell The building block of a crystal which, when repeated by pure translations, can generate the entire crystal. The unit cell is a parallelepiped whose edges are defined by three non-coplanar vectors conventionally called *a, b, c*. Unit cells within the crystal are separated by an integral number of steps in these vectors.

van der Waals force A weak force existing between nearby nonbonded atoms. The repulsive van der Waals force arises from the Pauli exclusion principle and resists atoms being pushed too close together; this is often modeled as being proportional to $(1/R^{12})$ where R is the center-to-center separation of two atoms. The attractive part of the force arises from dipole–dipole interactions, in which quantum mechanical fluctuations in the electron cloud around one atom induce a transient dipole in a neighboring atom. Often modeled as proportional to $-(1/R^6)$.

Vapor diffusion See *hanging drop method*.

Variance A measure of the dispersion around the mean of independent measurements of a quantity. If x_i are the n measurements of a random variable with mean $<x>$, the variance v is given by

$$v = \frac{1}{n-1} \sum_{1}^{n} (x_j - \langle x \rangle)^2$$

Weighting factor A multiplicative factor used in refinement and statistical analyses that emphasizes terms that are known with high confidence. Often equal to (1/variance) of the measurement involved.

Wiggler In synchrotrons, an *insertion device* similar to an *undulator*, but producing a white, rather than a linelike, spectrum. Wigglers contain fewer dipole elements than an undulator.

Wilson plot A plot that allows one to determine the effective overall *temperature factor* for a diffraction data set. It is based on the relation $I_{obs} = I_{calc} \exp(-2B \sin^2 \theta / \lambda^2)$. If one averages over all the reflections in a fairly thin spherical shell in reciprocal space, one can rewrite the equation above for each shell as

$$\ln \left[\frac{\langle I(h) \rangle_{shell}}{\displaystyle\sum_{h}^{in\ shell} f_j^2} \right] = \ln (\text{constant}) - \frac{2B \sin^2 \theta_{shell}}{\lambda^2}$$

A plot of $\ln(<I(h)>/\Sigma f^2)$ versus $\sin^2 \theta_{shell}/\lambda^2$ should therefore give a straight line. B can be obtained from the slope of this line.

Index

Page numbers followed by an *f* indicate figures.